PEAR-SHAPED
NUCLEI

PEAR-SHAPED
NUCLEI

Suresh C Pancholi

Inter University Accelerator Centre, New Delhi, India

World Scientific

EW JERSEY · LONDON · SINGAPORE · BEIJING · SHANGHAI · HONG KONG · TAIPEI · CHENNAI · TOKYO

Published by

World Scientific Publishing Co. Pte. Ltd.

5 Toh Tuck Link, Singapore 596224

USA office: 27 Warren Street, Suite 401-402, Hackensack, NJ 07601

UK office: 57 Shelton Street, Covent Garden, London WC2H 9HE

Library of Congress Control Number: 2020006491

British Library Cataloguing-in-Publication Data
A catalogue record for this book is available from the British Library.

PEAR-SHAPED NUCLEI

ISBN 978-981-121-759-3 (hardcover)
ISBN 978-981-121-760-9 (ebook for institutions)
ISBN 978-981-121-761-6 (ebook for individuals)

For any available supplementary material, please visit
https://www.worldscientific.com/worldscibooks/10.1142/11754#t=suppl

Typeset by Stallion Press
Email: enquiries@stallionpress.com

Preface

In the last several decades, the study of nuclear shapes has gained prime importance. A large number of investigations, both theoretical and experimental, have led to the discovery of a rich variety of nuclear shapes like, the basic spherical, deformed, superdeformed, triaxial, shape coexistence, reflection asymmetric (pear-shape) and other exotic shapes. Apart from common nuclear structural properties, each of the mentioned shapes manifests properties associated with its specific shape. It is interesting to note that most deformed nuclei are prolate deformed. In this monograph, attention is paid to pear-shaped nuclei.

This monograph is intended for new researchers or young researchers in nuclear structure physics. A pedagogical approach has therefore been kept in mind. It is well-known that while trying to learn about a topic for research, the researcher has to spend a fairly large amount of time in undertaking literature search. The author has made an attempt to provide the readers with an up-to-date subject knowledge in a consolidated and focused manner. One should acknowledge the immense contribution from the available topical review articles that have appeared in literature from time-to-time. The contents in this monograph, as expected will also be very useful for the teaching of an advanced nuclear physics class in universities and institutions. Pre-requisites for a proper appreciation of the monograph are basic and advanced-level nuclear physics courses, knowledge of basic experimental methodology and techniques.

Chapter 1 is devoted to the introduction of the topic, like the pioneering experimental research work which led to the discovery of pear-shaped nuclei, nuclear structure of the observed positive and negative parity bands in even-even actinide nuclei, nuclear shapes, microscopic origin of octupole correlations, nuclear potential energies and the simplex quantum numbers. In Chapter 2, the information on near ground state quadrupole shapes, evolution of octupole shapes in ground states, ground state spins and parities is presented and the available information from nuclear charge radii. Chapters 3 and 4 form the main text body of the monograph. It should be mentioned here that for the discussion on pear shapes, only a representative set of even-even and odd-mass nuclei have been chosen in the actinide and the lanthanide regions. Although, octupole shapes have also been investigated in other nuclei in both the mass regions and in the lighter mass regions, like A $\sim$ 80. Chapter 3 gives an in-depth discussion on excitation energy systematic of nuclear energy levels, energy splitting and alternating parity bands, displacement energy or parity splitting, octupole vibrational and deformed nuclei and rotational properties of the even-even nuclei. In this chapter, also considered are some similar properties for the odd-mass nuclei. In Chapter 4, the electromagnetic properties of transitions between the observed energy levels are discussed. Specifically discussed are, the B(E1)/B(E2) ratios, B(E1), B(E2), B(E3) transition rates and the electric dipole, quadrupole and octupole moments in these nuclei.

The last chapter, Chapter 5, is devoted to a summary and review of the experimental findings mentioned in Chapters 3 and 4. Presently, the pear-shaped nuclei have acquired importance of a fundamental nature. The *atomic* electric dipole moments in such odd mass nuclei are expected to be highly enhanced due to probable contributions from the CP-violating interactions. Amongst such CP-violations discovered are the decay of particles, like the K^0, B^0 and very recently D^0 mesons. The CP-violations, when confirmed in the odd mass pear-shaped nuclei, lead to the possibility of a partial solution to the matter-antimatter asymmetry problem in the universe.

This monograph is based on the numerous painstaking investigations, both theoretical and experimental, over several decades by a large number of physicists contributing to the expansion and addition of new knowledge on pear-shaped nuclei. I have tried to do justice to them by incorporating their works, but surely, I must offer my apologies for any shortcomings. This monograph is mainly based on experimental findings.

Most of the work on this monograph has been carried out at the Inter University Accelerator Centre (IUAC), New Delhi. I very gratefully acknowledge the excellent work independence, ambience and infrastructure support that I enjoyed at IUAC. I am especially indebted to the former Directors, Professors Amit Roy and D. Kanjilal, and the present Director Professor A.C. Pandey for their constant encouragement and support for this work. I am grateful to Professor P.A. Butler FRS for his advice and help during the course of writing this monograph. I am thankful to Professors R. Palit, N. Panchapakesan and H. J. Wollersheim for critical reading of portions of the manuscript and for useful comments. I enjoyed numerous discussions with my colleagues Drs. S. Muralithar and R.P. Singh at IUAC. I thank Ms. Indu Bala for doing some calculations and helping me in various aspects in the preparation of the manuscript. My special thanks to Ms. Anupriya Sharma for help in creating the figures.

The University of Delhi is where I did interactional teaching and research. I express my gratitude to this premier University which gave me a chance to inculcate free academic thinking. This University is also my *alma mater*. I would like to extend special thanks to my colleagues Professor G. K. Mehta, Professor A. K. Jain, Professor R. K. Bhandari and Professor R. G. Sharma for their support and advice.

Last but not the least, I am thankful and grateful to my family — my wife Rani, daughter Bela, son Vineet, daughter-in-law Ranju and grand-daughter Meghna for their unflinching support, patience, tolerance and perseverance during the course of this academic endeavor.

Suresh C. Pancholi
New Delhi

Contents

Chapter 1

The Pear-Shaped Nuclei

1.1. Introduction

Pioneering experimental works on the topic in this monograph were done by a group from University of California, Berkeley [1–3]. During their investigations, it was established that in the α-decay of several even-even nuclei in the actinide region, in the daughter nuclei, further decay by γ-ray transitions populated the sequence of low-lying energy states with spin and parity 0^+, 2^+, 4^+. These sequences were interpreted as rotational bands in terms of the Bohr–Mottelson unified nuclear model [4, 5]. In addition to the positive parity even spin rotational bands, a low energy state was discovered in each of these nuclei which did not form part of the already found rotational band. The α-decay of ^{228}Th [1] populated a state at then measured energy of 217 keV in the daughter nucleus ^{224}Ra which further decayed by a γ-ray of $E_\gamma = 137$ keV to the 2^+ state and another γ-ray of $E_\gamma = 212$ keV to the 0^+ state. This type of gamma decay pattern was found in all the nuclei studied. These two low-energy γ-rays were found to be of $E1$ character on the basis of total transition intensity balance at the 217 keV level. This intensity balance required that the total internal conversion coefficients of these transitions be much less than unity. This is true only if these transitions are of $E1$ multipolarity, forming the basis of assigning the 217 keV level spin and parity of 1^- (see Fig. 1.1). This spin and parity assignment was further confirmed by alpha-gamma angular correlation measurements in the α-decay of ^{228}Th and in ^{226}Th and ^{230}U [2].

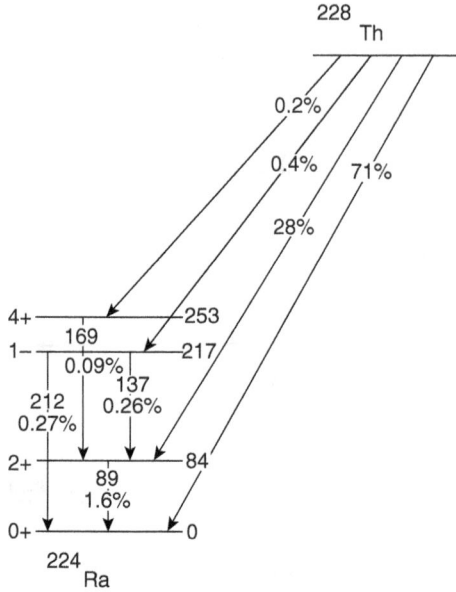

Fig. 1.1. Decay scheme of ^{228}Th [1] showing α-decay followed by γ-decay populating the positive parity states and the negative parity state 1^- at then measured energy of 217 keV in the daughter nucleus ^{224}Ra. Figure adopted from [1].

The question then arose as to what is the type of configuration of the 1^- states in these even-even nuclei which are low-lying and close in energy to the 2^+ and the 4^+ rotational states. A suggestion was made by R. Christy as communicated by A. Bohr mentioned in [3] that this "odd spin and odd parity level may have the same intrinsic structure as the ground state and represents a collective distortion in which the nucleus is pear-shaped."

Many years after the above mentioned experimental works, detailed high spin gamma-ray spectroscopic investigations in $N = 130$ ^{218}Ra nucleus [6] revealed the existence of even spin positive parity rotational band up to 14^+ and odd-spin odd parity rotational band to 17^- connected by enhanced $E1$ transitions. After this, experimental gamma-ray spectroscopic studies have been done, as cited in the literature, in many even-even, odd-N, odd-Z and odd-odd nuclei in the actinide and the lanthanide regions for the investigation of octupole shape distortions in nuclei. The derived level schemes from these investigations exhibited some common peculiarities. In even-even nuclei, like ^{220}Ra, ^{222}Th [7] etc., population of a positive parity ground state rotational band and a negative parity rotational band was shown. The neighboring opposite parity states in these bands are connected by enhanced $E1$ transitions. Initially at low spins, these bands have energy separation which gradually disappear with increasing spin and after a certain spin, the two bands form a single interleaved alternating parity band. In odd-N nuclei, like ^{223}Th, ^{225}Th [8, 9] etc., two pairs of rotational bands are found. Each pair consists of a positive parity and a negative parity band with the neighboring opposite parity states connected by enhanced $E1$ transitions. The states in the positive parity band (negative parity band) of one pair and the states in the negative parity band (positive parity band) in the other pair have same spin sequences but of opposite parity. Pairs of states with same spin but opposite parity form the parity doublet states and are nearly energy degenerate. In many such nuclei, electromagnetic decay properties of energy levels, like, reduced electromagnetic transition probabilities of electric dipole ($E1$), electric quadrupole ($E2$) transitions and in some cases involving magnetic dipole ($M1$) transitions, have been investigated. Also investigated are the electric dipole moments, electric quadrupole moments and electric octupole ($E3$) moments. These have been studied both experimentally and theoretically. In this monograph, these and other properties will be discussed in detail and their implications for the octupole (pear) shapes will be emphasized.

Excellent review articles on the subject of this monograph are available in the literature [10–12].

1.2. Nuclear Structure of Positive and Negative Parity Bands

Let us look into the suggestion of R. Christy cited above at least partially in the sense that the positive and the negative parity bands have the same nuclear structure. As mentioned above in Sec. 1.1 and discussed later in Chapter 3, Sec. 3.3, in several of the even-even Ra, Th and the lanthanide nuclei, after a certain initial spin, the positive and the negative parity bands form interleaved bands exhibiting an identical $I(I + 1)$ rotational pattern. The latter would mean that the rotational motion in the two bands is characterized by identical moment of inertia. This is the first test that the two bands have the same intrinsic structure.

The neighboring states in the negative parity and the positive parity ground state bands in these nuclei are interconnected by strong $E1$ transitions. The $E1$ transition rates are very sensitive to the nuclear structure of both states involved in the transition and the transition operator. Let us investigate the ratio of the reduced $E1$ transition probabilities from a negative parity state to two states in the positive parity band e.g., from 1^- to the 2^+ and the 0^+, 3^- to 4^+ and 2^+, etc. The ground state band has $K = 0$ (K denotes the projection of angular momentum I on the nuclear symmetry axis). We would like to know if the negative parity band also has $K = 0$ (same structure as the ground state band) or does it mix with $K = 1$ band due to Coriolis coupling. For small Coriolis mixing, one can write [13–15]:

$$\frac{B(E1; I^- \to (I+1)^+)}{B(E1; I^- \to (I-1)^+)} = \frac{I+1}{I} \left| \frac{1 - I \cdot z}{1 + (I+1) \cdot z} \right|^2 \quad (1.1)$$

where

$$z = \sqrt{2}\varepsilon_0 \frac{\langle 0^+ | M(1,-1) | 1^- \rangle}{\langle 0^+ | M(1,0) | 0^- \rangle} \quad (1.2)$$

Here the matrix elements of the $E1$ operator are $\langle 0^+ M(1,-K) \mid K^- \rangle$ and the $K^\pi = 1^-$ component in the wave function has the amplitude

$$c(I) = -\sqrt{2}\varepsilon_0 \sqrt{I(I+1)} \quad (1.3)$$

The $B(E1)$ ratio is sensitive to angular momentum, the structure of the nucleus involved and Coriolis mixing, if any, between the $K^\pi = 0^-$ negative parity band and the $K^\pi = 1^-$ band.

In Tables 1.1 and 1.2, in the last column, are presented the calculated values of the quantity $[\{I/(I+1)\}B(E1, I^- \to (I+1)^+)/B(E1, I^- \to (I-1)^+)]^{1/2}$ from the experimental data. Plots of

Table 1.1. Experimental branching ratios of $E1$ transitions from 1^- to 2^+ and 0^+ states and 3^- state to 4^+ and 2^+ states, from $K = 0^-$ band to $K = 0^+$ ground state band, for even-even Ra and Th nuclei. Experimental data on, spins and parities of levels I^π, level energy E(keV), gamma-ray transition energy E_γ (keV) from $I_i^\pi \to I_f^\pi$ and the relative gamma-ray intensity I_γ, have been taken from [17].

Nucleus	Initial level			Transition			$\sqrt{\frac{I}{I+1}\frac{B(E1;I^-\to(I+1)^+)}{B(E1;I^-\to(I-1)^+)}}$
	E(keV)	I^π	K	$I_i^\pi \to I_f^\pi$	E_γ(keV)	Rel. I_γ	
^{220}Ra	412.98	1^-	0	$1^- \to 2^+$	234.5	72(20)	1.4(2)
				$1^- \to 0^+$	413.0	100(13)	
^{222}Ra	242.11	1^-	0	$1^- \to 2^+$	131.00	32.1(16)	1.00(3)
				$1^- \to 0^+$	242.11	100(5)	
^{224}Ra	215.98	1^-	0	$1^- \to 2^+$	131.61	51.4(6)	1.06
				$1^- \to 0^+$	215.98	100.0(4)	
^{226}Ra	253.73	1^-	0	$1^- \to 2^+$	186.05	73(4)	0.96(4)[a]
				$1^- \to 0^+$	253.73	100(7)	
	321.54	3^-	0	$3^- \to 4^+$	110.00	10(3)	0.96(15)[a]
				$3^- \to 2^+$	253.9	100(11)	
^{228}Ra	474.18	1^-	0	$1^- \to 2^+$	410.40	82(4)	0.79(7)
				$1^- \to 0^+$	474.0	100(19)	
	537.50	3^-	0	$3^- \to 4^+$	332.91	25.1(16)	0.73(5)
				$3^- \to 2^+$	473.7	100(11)	
^{224}Th	251.0	1^-	0	$1^- \to 2^+$	152.9	50(13)	1.0(2)
				$1^- \to 0^+$	246	100(25)	
^{226}Th	230.37	1^-	0	$1^- \to 2^+$	158.18	60(5)	0.96(4)
				$1^- \to 0^+$	230.37	100(5)	
	307.5	3^-	0	$3^- \to 4^+$	81.0	4.1(10)	0.86(11)
				$3^- \to 2^+$	253.3	100(7)	
^{228}Th	328.02	1^-	0	$1^- \to 2^+$	270.24	100.0(16)	1.00(1)
				$1^- \to 0^+$	328.02	88.3(7)	
	396.09	3^-	0	$3 \to 4^+$	209.25	34.2(6)	1.04(1)
				$3^- \to 2^+$	338.32	100.0(17)	
^{230}Th	508.15	1^-	0	$1^- \to 2^+$	454.92	100(3)	1.07(3)
				$1^- \to 0^+$	508.15	60(3)	
	571.75	3^-	0	$3^- \to 4^+$	397.62	87(4)	1.20(4)
				$3^- \to 2^+$	518.54	100(5)	

[a]See also Ref. [19].

these values as a function of neutron number N are also shown in Fig. 1.2. It can be seen in the top two panels that this quantity is in agreement with that expected ($\sim$1) for a pure $K^\pi = 0^-$ configuration of the negative parity bands in ^{222}Ra, ^{224}Ra, ^{226}Ra and ^{224}Th, ^{226}Th,

Table 1.2. Experimental branching ratios of $E1$ transitions from 1^- to 2^+ and 0^+ states and 3^- state to 4^+ and 2^+ states, from $K = 0^-$ band to $K = 0^+$ ground state band, for even-even $^{140-148}$Ba nuclei. Experimental data on, spins and parities of levels I^π, level energy E(keV), gamma-ray transition energy E_γ(keV) from $I_i^\pi \to I_f^\pi$ and the relative gamma-ray intensity I_γ, have been taken from [17].

	Initial level			Transition			
Nucleus	E(keV)	I^π	K	$I_i^\pi \to I_f^\pi$	E_γ(keV)	Rel. I_γ	$\sqrt{\frac{I}{I+1}\frac{B(E1;I^-\to(I+1)^+)}{B(E1;I^-\to(I-1)^+)}}$
^{140}Ba	1802.90	3^-	0	$3^- \to 4^+$	672.1	25.4(13)	1.04(4)
				$3^- \to 2^+$	1200.25	100(5)	
^{142}Ba	1292.20	3^-	0	$3^- \to 4^+$	457.26	20(2)	1.12(10)
				$3^- \to 2^+$	932.82	100(15)	
^{144}Ba	758.94	1^-	0	$1^- \to 2^+$	559.57	98.1(25)	1.10(2)
				$1^- \to 0^+$	758.96	100(3)	
	838.37	3^-	0	$3^- \to 4^+$	308.23	14.6(7)	0.98(2)
				$3^- \to 2^+$	638.99	100.0(19)	
^{146}Ba	738.82	1^-	0	$1^- \to 2^+$	557.70	100.0(15)	1.88(3)
				$1^- \to 0^+$	738.86	32.9(8)	
	821.10	3^-	0	$3^- \to 4^+$	307.42	94.1(22)	2.52(5)
				$3^- \to 2^+$	640.10	100(3)	
^{148}Ba	687.2	1^-	0	$1^- \to 2^+$	545.5	87	0.93
				$1^- \to 0^+$	687.2	100	
	775.0	3^-	0	$3^- \to 4^+$	351.9	21	0.95
				$3^- \to 2^+$	633.2	100	

^{228}Th nuclei. However, in ^{230}Th ($N = 140$), for decay from the 3^- state, there may be a small $K^\pi = 1^-$ admixture. In the $N = 132$ ^{220}Ra nucleus, there is a large deviation from the expected value for $K^\pi = 0^-$ configuration for decay from the 1^- state in the negative parity band. The energy ratios $E(4^+)/E(2^+) = 2.29$ for this nucleus indicate its vibrational character. The validity of the $B(E1)$ ratio relation is then in question since it is for interconnecting transitions between the two rotational bands.

For the even-even Ba nuclei, the above mentioned quantity is plotted as a function of neutron number N in the bottom panel in Fig. 1.2. In ^{140}Ba, ^{142}Ba, ^{144}Ba and ^{148}Ba nuclei, this quantity is found to be in agreement with the expected value for pure $K^\pi = 0^-$ configuration for decay from the 3^- states in the negative parity

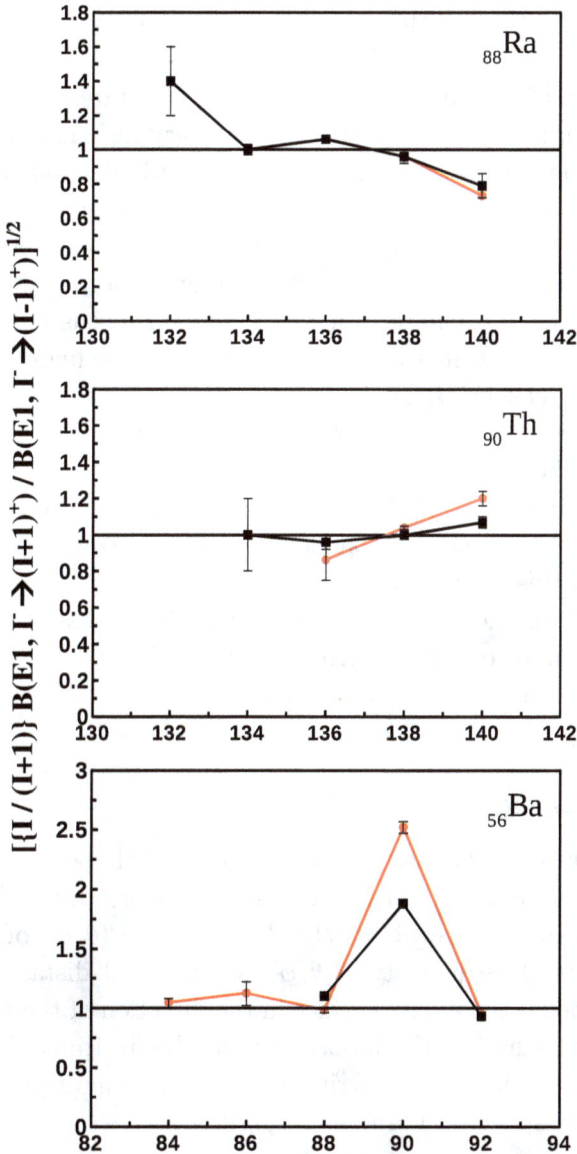

Fig. 1.2. Plots of the quantity $[\{I/(I+1)\}B(E1, I^- \to (I+1)^+)/B(E1, I^- \to (I-1)^+)]^{1/2}$ versus neutron number N for some even-even Ra, Th and Ba nuclei. This quantity is defined and explained in the text. The transition from 1^- and the 3^- states are denoted respectively in the above panels by (■) and (●). See Tables 1.1 and 1.2 for numerical data.

bands. The nucleus ^{146}Ba shows an unusual behavior (see also [16]). The quantity $[\{I/(I+1)\}B(E1, I^- \to (I+1)^+)/B(E1, I^- \to (I-1)^+)]^{1/2}$ for $E1$ decays from the 1^- and 3^- states to the positive parity ground state rotational bands are very high in comparison to unity. The energy ratio $E(4^+)/E(2^+) = 2.83$ [17] and deformation parameter $\beta_2 = 0.218$ [18] for this nucleus suggest ^{146}Ba to be having a rotational behavior. It should be mentioned here that the electric dipole moment D_0 in ^{146}Ba is an order of magnitude smaller than in ^{144}Ba (see Chapter 4, for a detailed discussion on electric dipole moments). But, this cannot account for the high values of the quantity $[\{I/(I+1)\}B(E1, I^- \to (I+1)^+)/B(E1, I^- \to (I-1)^+)]^{1/2}$ in this nucleus since a similar D_0 cancellation effect in ^{224}Ra does not affect the above quantity (see Fig. 1.2, top panel). Whether higher K-mixings should be considered is a question to be answered. It appears that a more plausible explanation is perhaps needed to understand this behavior.

Also, since the low lying negative parity states in these even-even nuclei have energies much lower than the expected energies of two quasi-particle states, these were interpreted as due to octupole vibrations of the nucleus about a spheriodal equilibrium shape [10].

1.3. Nuclear Shapes

When a nucleus deviates from the spherical shape and assumes a deformed shape at ground state, its shape can be described in the spherical system of coordinates by the function $R(\theta, \phi)$, of the surface of a nucleus. The quantity $R(\theta, \phi)$ is the radial distance from the centre of the nucleus to its surface in the direction of the polar angles θ and ϕ, measured in the laboratory coordinate frame. The general shape of the nucleus can be written in terms of a multipole expansion (sum) of the spherical harmonics $Y_{\lambda\mu}(\theta, \phi)$:

$$R(\theta, \phi) = R_0 \cdot \left[1 + \sum_{\lambda=0}^{\infty} \sum_{\mu=-\lambda}^{\lambda} \alpha_{\lambda\mu} \cdot Y_{\lambda\mu}(\theta, \phi) \right] \qquad (1.4)$$

where R_0 is the radius of the sphere of the same volume and $\alpha_{\lambda\mu}$ are the deformation parameters characterizing the shape of the deformed nucleus. Imposing the condition that the centre of mass is to be the same as the origin of the body-fixed frame and the nuclear shape is axially symmetric ($\mu = 0$), we then have

$$\beta_\lambda = \alpha_{\lambda 0}, \quad \lambda = 1, 2, 3, \ldots.$$

The experimental information tell us that only a few lowest multipolarities, like $\lambda = 2, 3, 4$ are important in the multipole expansion of the nuclear shape. For $\lambda = 0$, the nuclear shape is spherical. The deformation for $\lambda = 1$ (for small deformations) corresponds to the displacement of the nucleus as a whole in space and, therefore, cannot give rise to nuclear excitations and can be ignored. The $\lambda = 2$ quadrupole deformation term is very important as there is predominance of axially symmetric ($\mu = 0$) and reflection symmetric *prolate* quadrupole deformed shaped nuclei (see Fig. 1 in [20]; see also [21]). The $\lambda = 3$ term with $\mu = 0$ describes an axially symmetric and reflection asymmetric nuclear shape like pears. This type of nuclear shape which occurs in certain mass regions as mentioned

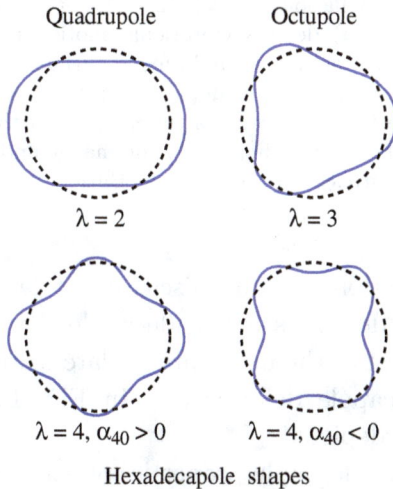

Quadrupole   Octupole

$\lambda = 2$ $\lambda = 3$

$\lambda = 4, \alpha_{40} > 0$ $\lambda = 4, \alpha_{40} < 0$

Hexadecapole shapes

Fig. 1.3. Quadrupole ($\lambda = 2$), octupole ($\lambda = 3$) and hexadecapole ($\lambda = 4$) nuclear shape deformations. The spherical shape ($\lambda = 0$) is shown in the figures as dashed circle. All the shapes are axially symmetric. While the quadrupole and the hexadecapole shapes are reflection symmetric, the octupole shape is reflection asymmetric. The figure is adopted from [22].

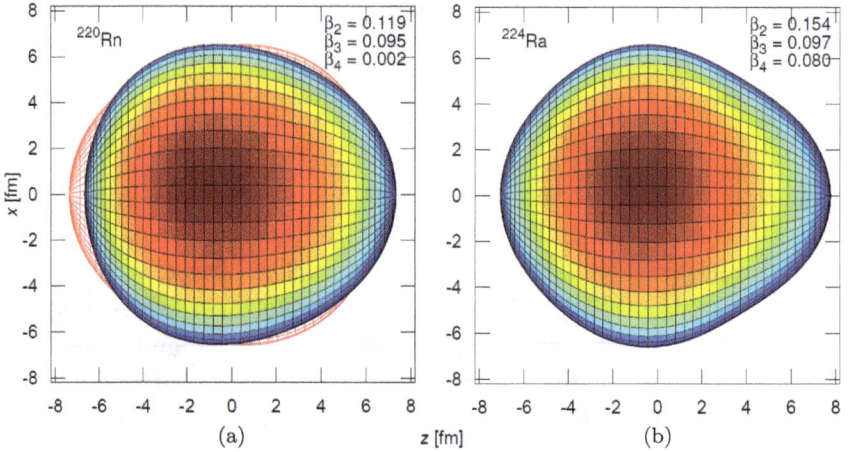

Fig. 1.4. In the figure, the shapes of ^{220}Rn and ^{224}Ra nuclei are graphically represented. The values of the deformation parameters β_2, β_3 and β_4 for the two nuclei are mentioned in the upper right corners of the left and right panels (**a**) and (**b**), respectively. These values are as given in [24]. The β_2 and β_3 values were extracted from the dependence of the measured quadrupole Q_2 and octupole Q_3 moments on the generalized nuclear shape [23, 25]. The theoretical values of β_4 were taken from the calculations in [23]. In ^{220}Rn, there is octupole vibration around $\beta_3 = 0$. Panel (**a**) depicts vibrational motion for this nucleus about symmetry between the surface shown and the red outline. In panel (**b**) is depicted the shape for ^{224}Ra nucleus — an example of permanent octupole deformation. The color scale in the figures represents the y-values of the surface. It may be mentioned here that the nuclear shape does not change under rotation about the z-axis. The figure is courtesy of Professor P.A. Butler.

in the next subsection, will be discussed at length in this book. The $\lambda = 4$ term describes a hexadecapole deformed nucleus. As already known, most of the nuclei are prolate deformed with a small additional hexadecapole deformation. In Fig. 1.3, the mentioned nuclear shapes are depicted.

In the present monograph, properties of octupole vibrational and octupole deformed nuclei will be presented. It will, therefore, be helpful to visualize the calculated shapes of these axially symmetric reflection asymmetric nuclei in some representative cases. In Fig. 1.4, the deduced shapes in ^{220}Rn and ^{224}Ra nuclei are shown. (See the figure caption for the details of shape deduction.)

1.4. Microscopic Origin of Octupole Correlations

The origin of octupole vibration or octupole deformation or to say, in general, octupole correlations, can be understood through the single particle level diagram shown in Fig. 1.5. This shows

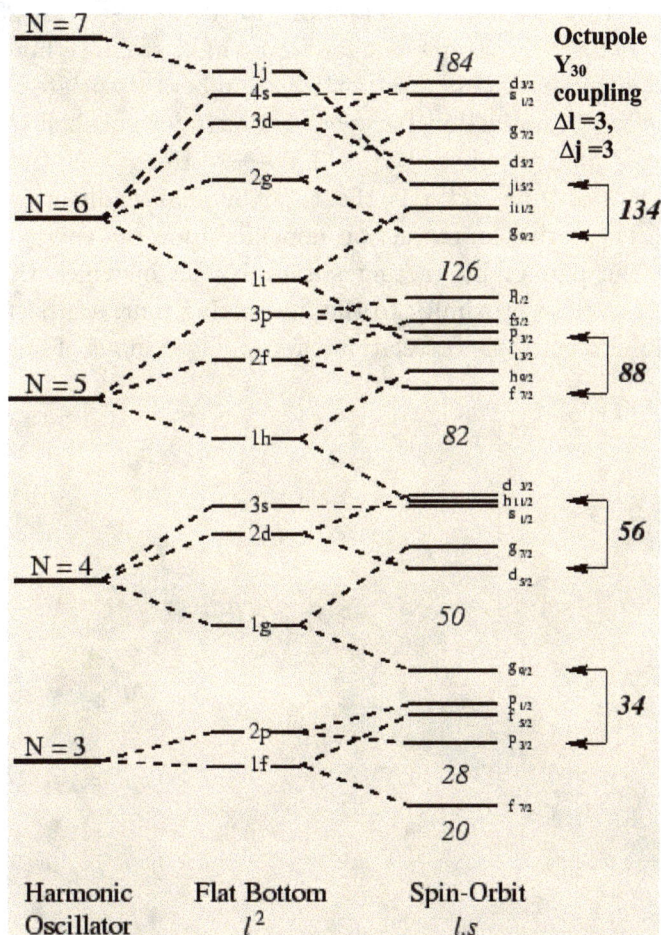

Fig. 1.5. Splitting and ordering of single particle levels for neutrons due to l^2 and $l.s$ terms added to the harmonic oscillator potential. The numbers in the middle of the level diagram on the right side denote magic numbers. The neutron and the proton level diagrams are quite similar. The numbers 34, 56, 88 and 134 on the extreme right are the particle numbers for the favored $\Delta l = \Delta j = 3$ octupole Y_{30} coupling. The interacting orbitals are also mentioned alongside the levels.

the splitting and ordering of single particle levels for neutrons due to the l^2 and the spin-orbit $l.s$ interaction terms added to the harmonic oscillator potential. The proton and the neutron single particle states are quite similar. Microscopically, a coupling between intrinsic states of opposite parity is produced by the long-ranged octupole-octupole residual interaction (for details, see [11, 23]). For normally-deformed systems not far from the β stability line, strong octupole coupling exists for particle numbers associated with a large $\Delta N = 1$ interaction between the intruder subshell (l, j) and the normal parity subshell $(l - 3, j - 3)$. Here, l is the orbital angular momentum and j is the total angular momentum of the particle. The octupole interaction depends upon the energy spacing between the pair of interacting states. Nuclei in which the Fermi surface is in close proximity to the interacting pair, are particularly susceptible to octupole correlation effects. The regions of nuclei with

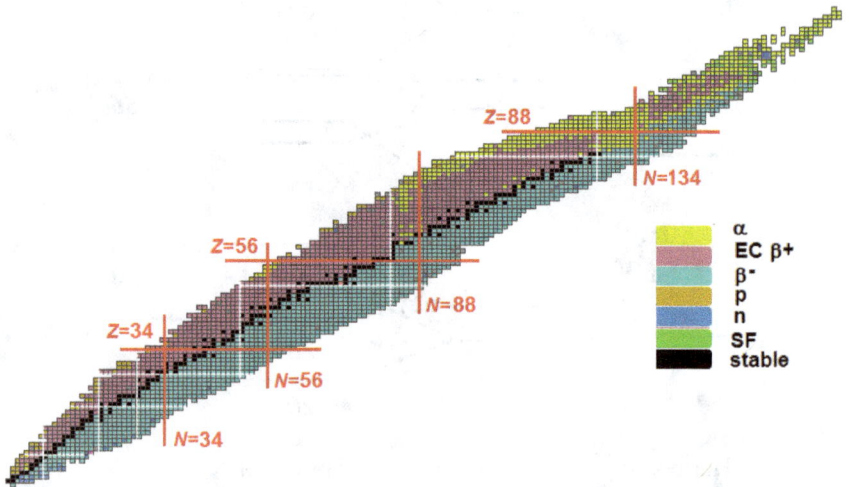

Fig. 1.6. The Z versus N chart of nuclides with the neutron and proton numbers (34, 56, 88 and 134) having strong octupole couplings marked in red. The white lines are the positions of the magic numbers. It is note worthy that strong octupole correlations are expected in all the N, Z regions just above the magic numbers. The stable isotopes are marked in black and the various radioactive decay modes are shown in different colors as given in the legend on the right. The figure is courtesy of Professor P.A. Butler.

strong octupole correlations correspond to particle numbers (neutron or proton) in close proximity to 34 (coupling between $1g_{9/2}$ and $2p_{3/2}$ orbitals), 56 (coupling between $1h_{11/2}$ and $2d_{5/2}$ orbitals), 88 (coupling between $1i_{13/2}$ and $2f_{7/2}$ orbitals) and 134 (coupling between $1j_{15/2}$ and $2g_{9/2}$ orbitals), that is, just above the closed shells (see Fig. 1.6). These particle numbers and the orbitals are mentioned on the right side of the shell model diagram. In general, the energy spacing between the octupole-driving $\Delta l = \Delta j = 3$ orbital pair decreases with increasing mass number. Strongest octupole correlations are, thus predicted and observed in $Z \sim 88$ and $N \sim 134$ actinide Ra-Th region.

1.5. Nuclear Potential Energies

The nuclear potential energy as a function of quadrupole (β_2), octupole (β_3) and hexadecapole (β_4) deformation parameters can be calculated by the consideration of modified oscillator, folded Yukawa or the Woods–Saxon single particle potentials [23] using the microscopic-macroscopic method. Here, we consider the nuclear potential energy (based on folded Yukawa potential) in ground state as a function of octupole deformation parameter β_3 for a quadrupole plus octupole deformed axially symmetric even-even nucleus [26] in the actinide mass region. In Fig. 1.7, the potential energy versus β_3 is considered in the below mentioned three situations. In the left most panel, the potential barrier reaches a minimum at $\beta_3 = 0$, the nucleus is then on an average reflection symmetric and undergoes octupole vibrations around $\beta_3 = 0$. The expected level structure will be as shown below the potential energy plot. In the right most panel plot of potential energy, there are two degenerate minima at $\pm\beta_3$ (corresponding to a reflection asymmetric shape and its mirror image). The potential barrier rises to infinity at $\beta_3 = 0$. In this situation, a rigid octupole shape ensues for the nucleus and the expected level sequence will be a perfectly interleaved even and odd spin rotational band with alternating parity, like, 0^+, 1^-, 2^+,..., as shown at the bottom of the right most potential energy plot. As it will be seen later, this is not achieved in actual nuclei. Let us consider the potential energy

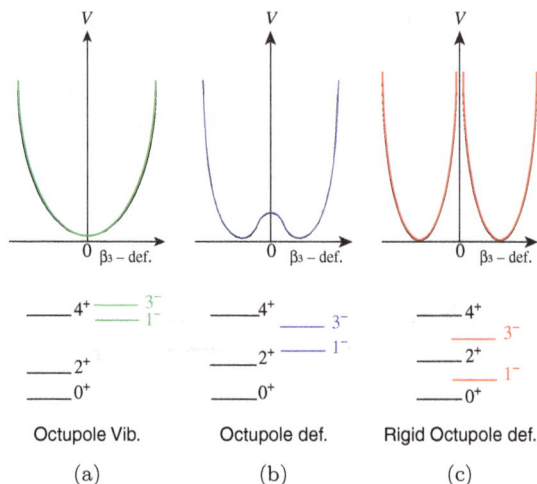

Fig. 1.7. Plots of potential energy V versus octupole deformation β_3 for different axially symmetric ($K = 0$) octupole shapes in even-even nuclei: octupole vibrational (left panel), rigid octupole deformed (right panel) and octupole deformed (or nuclei with strong octupole correlations) (middle panel) nuclei. The figure is adopted from [27]. See also [12].

plot in the middle panel which is intermediate between the other two described above and is closer to the situation that prevails in actual nuclei even with strong octupole correlations. At $\beta_3 = 0$, there is a small (<0.5 MeV) potential barrier. The nucleus can tunnel through it to its mirror image shape resulting in energy displaced positive parity even spin and negative parity odd spin bands. The former band is energetically favored. A representative level sequence in such a situation is shown at the bottom of the potential plot.

Below, in the following two figures, are shown the actual level schemes found in some specific nuclei. Figure 1.8, on the left is the level scheme of ^{220}Rn nucleus. For this nucleus, $E(4+)/E(2+) = 2.21$ and $\beta_2 = 0.127(22)$ [28]. It can, therefore, be considered as a near quadrupole vibrational nucleus. As mentioned later in Chapter 3, Sec. 3.5.3, this nucleus is also octupole vibrational. The 1^- level here is at an excitation energy of 645.44 keV as compared to the 0^+ ground state. This then is an example of nuclear potential energy situation as in Fig. 1.7(a). The level scheme of ^{224}Ra ($N = 136$) nucleus is shown

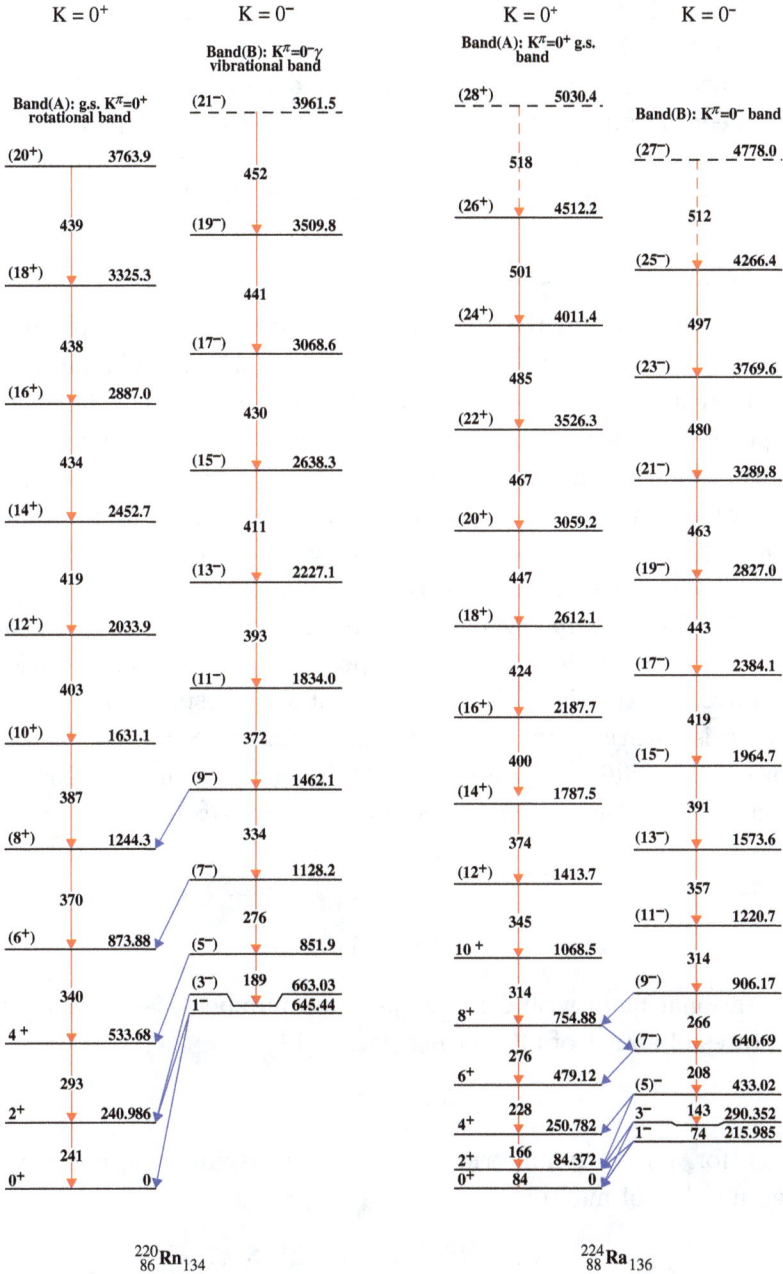

Fig. 1.8. Level schemes of ^{220}Rn and ^{224}Ra [17 and references therein].

on the right in this figure. In this nucleus, $E(4+)/E(2+) = 2.97$ and $\beta_2 = 0.1790$ (12) [28]. It is a quadrupole prolate deformed nucleus. As mentioned later in Sec. 4.6, it is amongst one of the few nuclei which have been found to be an octupole deformed pear-shaped nucleus. Notice that the 1^- level here is at low excitation energy of 215.98 keV. This is an example of the situation in Fig. 1.7(b).

1.6. The Simplex Quantum Number, s

It is well-known that the ground-state band in an even-even nucleus has the level sequence with $J^\pi = 0^+$, 2^+, 4^+, 6^+, ... The nucleus is axially symmetric ($\mu = 0$) and it is reflection *symmetric*. The nuclear shape of such a nucleus is described by only the even multipole deformation components $\lambda = 2, 4, \ldots$. The nuclear wave function in this case is invariant with respect to both R, a rotation by 180° about an axis perpendicular to its symmetry axis and the space inversion P (parity operator). In the case of an axially symmetric ($\mu = 0$) but reflection *asymmetric* nucleus, the shape is described by the odd multipole deformations, like $\lambda = 3$ (pear shape). For such a nucleus, the wave function is no longer invariant with respect to R, nor P. But, it is invariant to the combined operation $S = PR^{-1}$, which represents a reflection in a plane containing the symmetry axis. The eigenvalue s of the S operator is called the simplex quantum number s [29 and references therein]. The square of the S operator is related to the total number of fermions A, by the relation,

$$S^2 = (-1)^A$$

A rotational band with simplex quantum number s is characterized by states of spin I of alternating parity [14]

$$p = s\,e^{-i\pi I}$$

Thus, for axially symmetric and reflection asymmetric nuclei with even number of nucleons, we have

$$s = +1, \quad I^p = 0^+,\ 1^-, 2^+, 3^-, \ldots$$
$$s = -1, \quad I^p = 0^-,\ 1^+, 2^-, 3^+, \ldots$$

and for nuclei with odd particle number, we have

$$s = +i, \quad I^p = 1/2^+, 3/2^-, 5/2^+, 7/2^-, \ldots$$
$$s = -i, \quad I^p = 1/2^-, 3/2^+, 5/2^-, 7/2^+, \ldots$$

Therefore, the rotational properties or rotational excitations of octupole deformed nuclei can be classified by this simplex quantum number s.

References

1. F. Asaro, F. Stephens, Jr. and I. Perlman, *Phys. Rev.* **92**, 1495 (1953).
2. F. Stephens, Jr., F. Asaro and I. Perlman, *Phys. Rev.* **96**, 1568 (1954).
3. F. Stephens, Jr., F. Asaro and I. Perlman, *Phys. Rev.* **100**, 1543 (1955).
4. A. Bohr and B.R. Mottelson, Kgl. Danske Videnskab, Selskab Mat.-fys. Medd. 27, No. 16 (1953).
5. A. Bohr, *Rotational States of Atomic Nuclei* (Ejnar Munksgaard, Copenhagen, 1954).
6. J. Fernández-Niello *et al.*, *Nucl. Phys. A* **391**, 221 (1982).
7. J.F. Smith *et al.*, *Phys. Rev. Lett.* **75**, 1050 (1995).
8. M. Dahlinger *et al.*, *Nucl. Phys. A* **484**, 337 (1988).
9. J.R. Hughes *et al.*, *Nucl. Phys. A* **512**, 275 (1990).
10. I. Ahmad and P.A. Butler, *Annu. Rev. Nucl. Part. Sci.* **43**, 71 (1993).
11. P. Butler and W. Nazarewicz, *Rev. Mod. Phys.* **68**, 349 (1996) and references therein.
12. P. Butler, *J. Phys. G: Nucl. Part. Phys.* **43**, 073002 (2016).
13. L. Kocbach and P. Vogel, *Phys. Lett. B* **32**, 434 (1970).
14. A. Bohr and B.R. Mottelson, *Nuclear Structure Vol.* 2, Reading, Massachusetts, W.A. Benjamin (1975).
15. P. Zeyen *et al.*, *Z. Phys. A* **328**, 399 (1987).
16. D. Kusnezov and F. Iachello, *Phys. Lett. B* **209**, 420 (1988).
17. Brookhaven National Data Center, ENSDF files; http://www.nndc.bnl.gov and references therein.
18. S. Raman *et al.*, *At. Data Nucl. Data Tables* **78**, 1 (2001).
19. B. Ackermann *et al.*, *Z. Phys. A* **355**, 151 (1996).
20. Stránský *et al.*, *J. Phys: Conf. Series* **322**, 012018 (2011).
21. D. Bonatsos, *Eur. Phys. J. A* **53**, 148 (2017).
22. P. Ring and P. Schuck, *The Nuclear Many-body Problem* (Springer, 1980).
23. W. Nazarewicz *et al.*, *Nucl. Phys. A* **429**, 269 (1984) and references therein.
24. L.P. Gaffney *et al.*, *Nature* **497**, 199 (2013).
25. G.A. Leander and Y.S. Chen, *Phys. Rev. C* **37**, 2744 (1988).
26. G.A. Leander *et al.*, *Nucl. Phys. A* **388**, 452 (1982).

27. L.P. Gaffney, Ph.D., Thesis, University of Liverpool (2012).
28. B. Pritychenko *et al.*, *At. Data Nucl. Data Tables* **107**, 1 (2016).
29. W. Nazarewicz and P. Olanders, *Nucl. Phys. A* **441**, 420 (1985) and references therein.

Chapter 2

Ground-state Nuclear Shape Deformations

2.1. Introduction

Quadrupole-octupole deformed shapes exist in the ground states and in the rotational states, that is, at high rotational frequencies or high spins in the lanthanide with $Z \sim 56$, $N \sim 90$ and light actinide with $Z \sim 88$, $N \sim 136$ nuclei and could also be present in some other light and very heavy nuclei [1–3 and references therein]. The ratio of excitation energies of the first 4^+ and 2^+ energy states in even-even nuclei is a general signature of whether a nucleus is vibrational, transitional or deformed near the ground state. This is considered as a function of neutron number in Sec. 2.2. In the next section (Sec. 2.3), a global theoretical approach to predict the ground state nuclear octupole shapes is discussed. A comparison of the experimental and the theoretically predicted (from nuclear model considerations) ground state spins and parities in odd-mass nuclei beyond Pb, revealed certain regional discripancies. This could be removed when in theoretical calculations an octupole shape was considered in addition to the prolate deformed shape. This aspect is detailed in Sec 2.4. A very interesting finding that the differences in nuclear charge radii could be an indicator of nuclear shapes, is discussed in the last Sec. 2.5.

2.2. Near Ground State Quadrupole Shapes

One of the most commonly accepted practice to learn about the near ground state shape of a nucleus whether it is vibrational, transitional

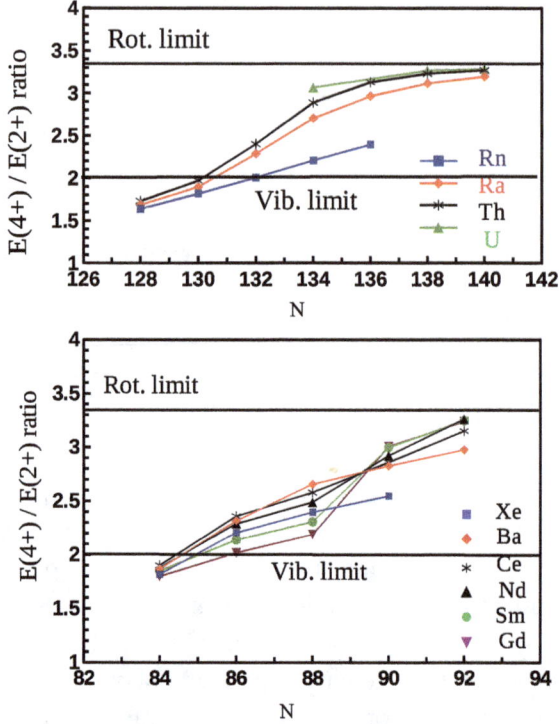

Fig. 2.1. Excitation energy ratios $E(4^+)/E(2^+)$ for the positive parity ground state band levels in the even-even Rn, Ra, Th and U (upper panel) and Xe, Ba, Ce, Nd, Sm and Gd (lower panel) isotopes as a function of neutron number N. The two horizontal lines in each of the panel are drawn at the vibrational and the rotational limits 2.00 and 3.33 respectively, of the $E(4^+)/E(2^+)$ energy ratios. The experimental data on energy levels are taken from [17 and references therein].

or deformed, is to determine the excitation energy ratio $E(4^+)/E(2^+)$ of the first 4^+ and the 2^+ energy levels of the ground state band in an even-even nucleus. Such ratios are plotted as a function of neutron number for the even-even lighter Rn, Ra, Th and U and Xe, Ba, Ce, Nd, Sm and Gd nuclei in the upper and the lower panels of Fig. 2.1 respectively. Also marked in these figures are the rotational $[E(4^+)/E(2^+) = 3.33]$ and the vibrational $[E(4^+)/E(2^+) = 2.0]$ limits of the ratios as horizontal lines. It is relevant to mention here that in most of the above mentioned nuclei in which octupole correlations have been found, they lie in the transitional region between the quadrupole vibrators and the deformed quadrupole rotors.

2.3. Evolution of Octupole Shape in Ground States

A large number of theoretical investigations have been done on the evolution of octupole nuclear shapes in ground states in the actinide and the lanthanide nuclei, as reported in the literature [3–16].

In the following, theoretical predictions with regard to the evolution of nuclear shapes found in [3] will be discussed. In this work, a global search for octupole shapes in the ground states was undertaken for all $Z \leq 106$ even-even nuclei located between two-proton and two-neutron drip lines within the framework of a number of covariant energy density functionals (CEDFs). In addition to the known regions of octupole deformed shapes, a new region of octupole deformation centered around $Z \sim 98$ and $N \sim 196$ was predicted by this theory. Very interesting results were obtained for nuclei with known octupole correlations, using the density dependent point coupling (DD-PC1) functionals. Calculations have been done to obtain the potential energy surfaces (PESs) in the quadrupole and octupole deformation parameters (β_2, β_3) plane, values of β_2, β_3 and a quantity ΔE^{oct}, as a function of neutron number N. The latter quantity has been defined as:

$$\Delta E^{\text{oct}} = E^{\text{oct}}(\beta_2, \beta_3) - E^{\text{quard}}(\beta_2', \beta_3' = 0),$$

where $E^{\text{oct}}(\beta_2, \beta_3)$ and $E^{\text{quard}}(\beta_2', \beta_3' = 0)$ are binding energies in the local minimum for quadrupole plus octupole shape and for quadrupole shape alone with no octupole deformation, respectively. In the calculation of the above two quantities, a variational approach was taken [18]. As a result, the quadrupole deformation obtained in the above two cases are different. The quantity $|\Delta E^{\text{oct}}|$ is then the gain in the binding energy due to octupole deformed shape. Small gain in binding energy, i.e. small $|\Delta E^{\text{oct}}|$ will be characteristic for soft PES which is typical for octupole vibrational nuclei. Large values of $|\Delta E^{\text{oct}}|$ correspond to well-pronounced local minima in the PES which will prefer stable octupole deformed shape. Therefore, the quantity $|\Delta E^{\text{oct}}|$ is an indicator of stability of octupole deformed shape.

In Figs. 2.2–2.4, the plots of potential energy surfaces for the Rn, Ra and Th isotopes in the (β_2, β_3) plane are shown [3].

In the ^{220}Rn, ^{222}Rn and ^{224}Rn isotopes investigated, no octupole deformation is predicted. For ^{222}Rn ($N = 136$) and ^{224}Rn ($N = 138$) isotopes, the PESs are soft in the β_3 direction (Fig. 2.2). In the equilibrium shape calculations in the microscopic-macroscopic approach [19] with Woods–Saxon potential, only ^{222}Rn and ^{224}Rn have finite β_3 deformation, $\beta_3 = 0.078$ and 0.060 respectively. The gain in the binding energy is small, $|\Delta E^{\text{oct}}| \sim 100$ keV. The experimentally determined values of the difference of aligned angular momentum Δi_x, at high spins in ^{218}Rn, ^{220}Rn and ^{222}Rn when plotted against the rotational frequency $\hbar\omega$, suggest that these Rn isotopes behave like octupole vibrators ($\Delta i_x \sim 3\,\hbar$) (see Chapter 3, Sec. 3.5.3 and [2]).

The potential energy surfaces for $^{218, 220, 222, 224, 226 \text{ and } 228}$Ra isotopes in the (β_2, β_3) plane are shown in Fig. 2.3. In ^{218}Ra and ^{220}Ra nuclei, weakly quadrupole deformed minima with octupole deformation $\beta_3 = 0$, are found. Following the PESs as a function of neutron number, one observes that the octupole minima have pronounced formation for ^{224}Ra ($N = 136$) and ^{226}Ra ($N = 138$) nuclei. With further increase in neutron number, the PESs become soft in the octupole direction. In the equilibrium shape calculations in the microscopic-macroscopic approach [19] with Woods–Saxon potential, only ^{222}Ra ($N = 134$), ^{224}Ra ($N = 136$) and ^{226}Ra ($N = 138$) have finite octupole deformation, $\beta_3 = 0.092$, 0.099 and 0.083, respectively. In Fig. 2.5, theoretical values of β_2, β_3 and ΔE^{oct} from [3] are plotted as a function of neutron number for the Ra and Th isotopes. Also plotted, are the experimental values of the excitation energy of the 1^- bandheads of the negative parity bands and the experimental values of β_2-deformation. It is found that the gain in binding energy, $|\Delta E^{\text{oct}}|$ is maximum for ($N = 136$) ^{224}Ra nucleus. The experimental and theoretical values of β_2-deformation are in good agreement for the Ra isotopes. For the ($N = 136$) ^{224}Ra nucleus, $E(1^-$ level$) = 215.98$ keV, which is the lowest in energy compared to the neighboring Ra isotopes.

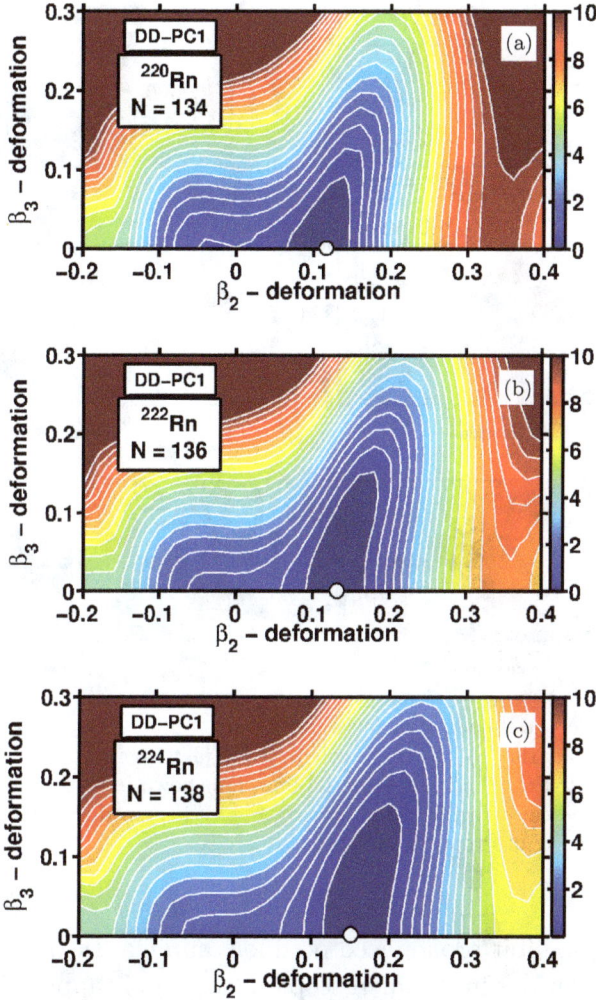

Fig. 2.2. Potential energy surfaces for ^{220}Rn, ^{222}Rn and ^{224}Rn isotopes in the (β_2, β_3) plane calculated in [3] with covariant energy density functionals DD-PC1. The equipotential lines are shown in steps of 0.5 MeV. The white circle indicates the global minimum. Figure is reproduced with permission from [3], courtesy of Professor Afanasjev.

The plots of the potential energy surfaces for ^{220,222,224,226,228,230}Th isotopes in the (β_2, β_3) plane are shown in Fig. 2.4. If the PESs are followed as a function of neutron number, these appear to have a behavior similar to the Ra isotopes. ^{226}Th ($N = 136$) and ^{228}Th

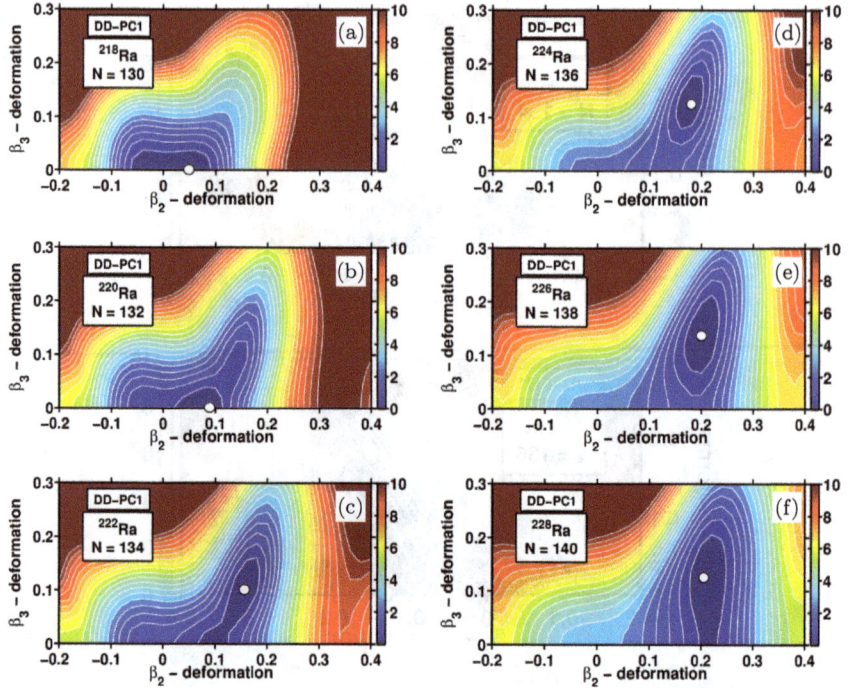

Fig. 2.3. Potential energy surfaces for ^{218}Ra, ^{220}Ra, ^{222}Ra, ^{224}Ra, ^{226}Ra and ^{228}Ra isotopes in the (β_2, β_3) plane calculated in [3] with covariant energy density functionals DD-PC1. The equipotential lines are shown in steps of 0.5 MeV. The white circle indicates the global minimum. Figure is reproduced with permission from [3], courtesy of Professor Afanasjev.

($N = 138$) exhibit pronounced octupole minima. In the equilibrium shape calculations in the microscopic-macroscopic approach [19] with Woods–Saxon potential, only ^{222}Th ($N = 132$), ^{224}Th ($N = 134$) and ^{226}Th ($N = 136$) have significant octupole deformation, $\beta_3 = 0.096$, 0.107 and 0.108, respectively. The gain in binding energy, $|\Delta E^{\text{oct}}|$ for ($N = 136$) ^{226}Th is not only the highest amongst the neighboring Th isotopes but is also much higher in comparison to that in $N = 136$ ^{224}Ra nucleus (Fig. 2.5, third panel from top on the right). The excitation energy $E(1^-$ level) $= 230.37$ keV of the 1^- bandhead of the negative parity band in ^{226}Th is lowest in energy compared to the neighboring Th isotopes. This means that ^{226}Th

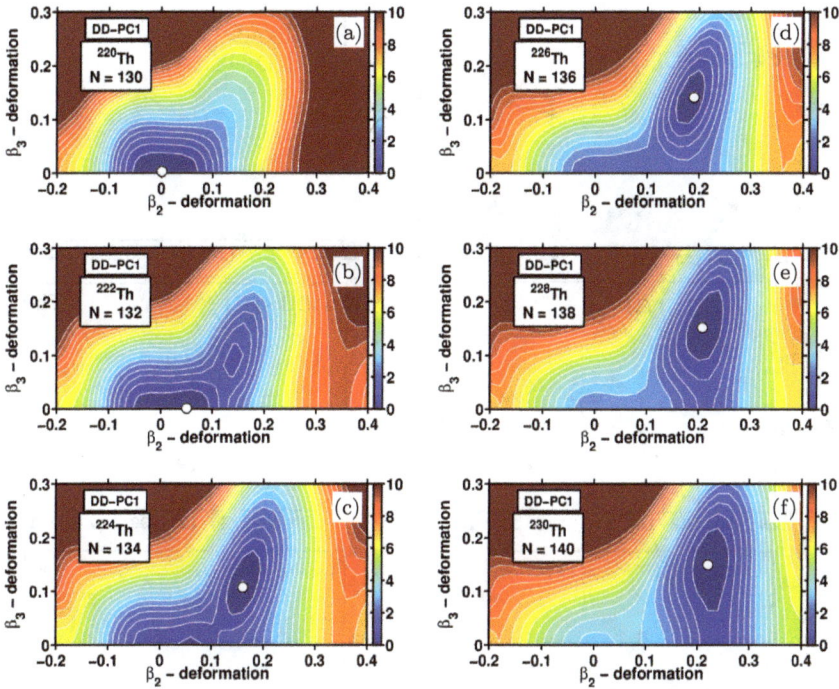

Fig. 2.4. Potential energy surfaces for ^{220}Th, ^{222}Th, ^{224}Th, ^{226}Th, ^{228}Th and ^{230}Th isotopes in the (β_2, β_3) plane calculated in [3] with covariant energy density functionals DD-PC1. The equipotential lines are shown in steps of 0.5 MeV. The white circle indicates the global minimum. Figure is reproduced with permission from [3], courtesy of Professor Afanasjev.

has a much higher stability for octupole deformed shape than that for ^{224}Ra.

In Fig. 2.6, theoretical values of β_2, β_3 and $\Delta E^{\rm oct}$ from [3] are plotted as a function of neutron number for the Ba isotopes. Also plotted, are the experimental values of the excitation energy of the 1^- bandheads of the negative parity bands and the experimental values of β_2-deformation. It is found that the gain in binding energy, $|\Delta E^{\rm oct}|$ is maximum for $(N = 90)$ ^{146}Ba nucleus. The experimental and theoretical values of β_2-deformation are in good agreement. The excitation energies of the 1^- and 3^- states in ^{144}Ba, ^{146}Ba and ^{148}Ba are shown in the bottom panel of the figure. In the equilibrium shape calculations in the microscopic-macroscopic approach [19] with

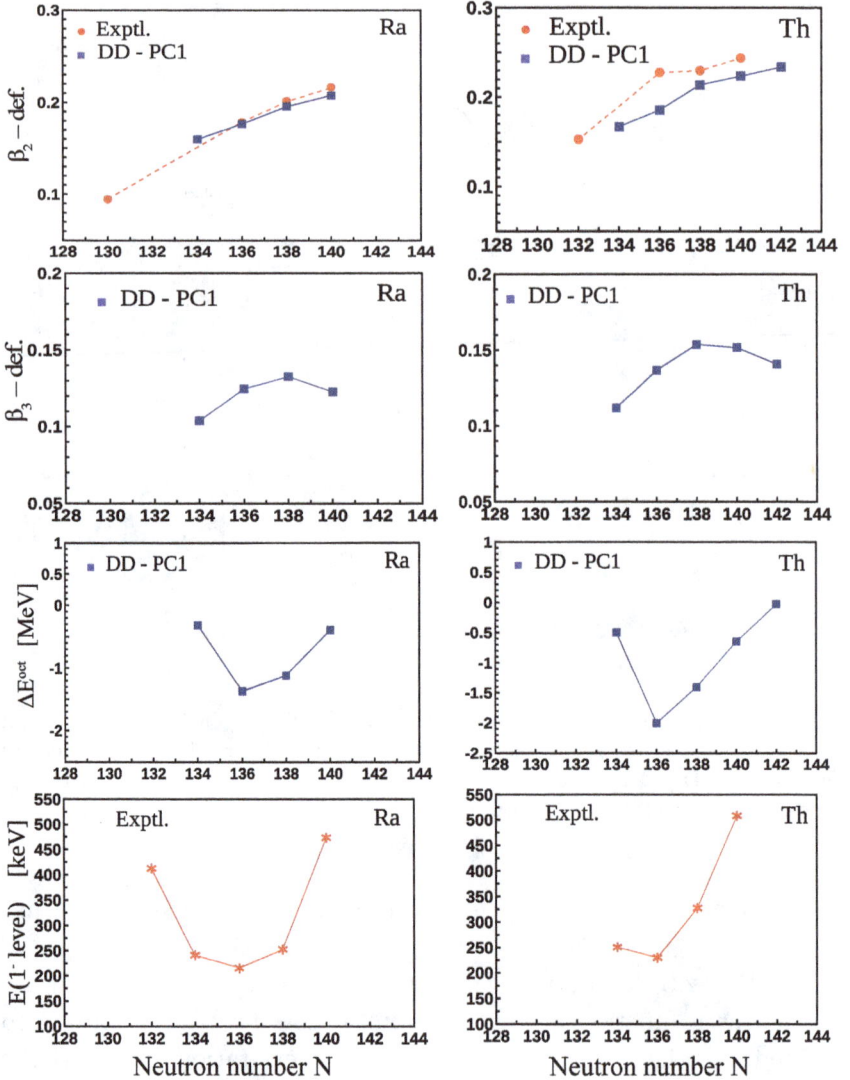

Fig. 2.5. Calculated [3] values of equilibrium quadrupole β_2 (■) (upper panels), octupole β_3 (■) (second panels from top) deformations and gain in binding energy due to octupole deformation ΔE^{oct} (■) (third panels from top) obtained in the relativistic Hartree–Bogoliubov (RHB) calculations with the DD-PC1 functional are plotted versus neutron number N for the Ra (left panels) and the Th (right panels) isotopes. The experimental values of β_2 (•) [20] are also shown for comparison in the upper panels. Also shown in the bottom panels are the energies of the 1^- states (∗) [17] in the Ra and Th isotopes.

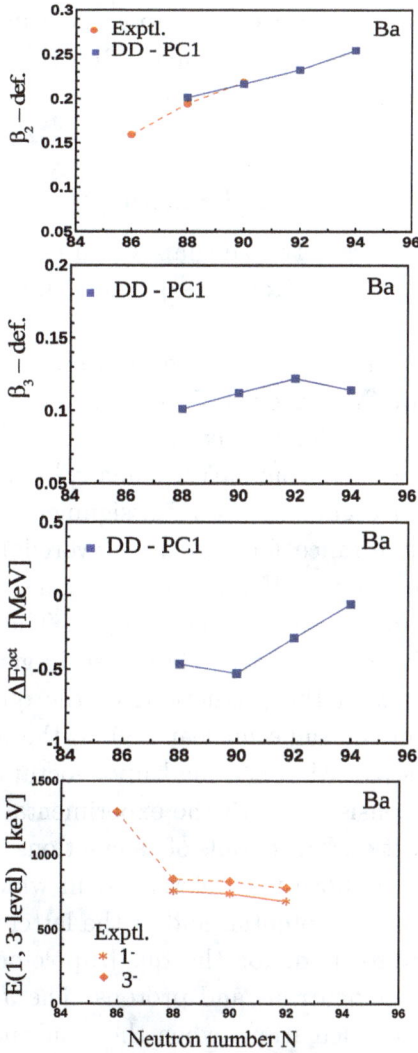

Fig. 2.6. Calculated [3] values of equilibrium quadrupole β_2 (■) (upper panel), octupole β_3 (■) (second panel from top) deformations and gain in binding energy due to octupole deformation ΔE^{oct} (■) (third panel from top) obtained in the relativistic Hartree–Bogoliubov (RHB) calculations with the DD-PC1 functional are plotted versus neutron number N for the Ba isotopes. The experimental values of β_2 (•) [20] are also shown for comparison in the upper panel. Shown in the bottom panel are the energies of the 1^- (∗) and the 3^- (♦) states [17] in the Ba isotopes.

Woods–Saxon potential, nonzero octupole deformation were found for ^{144}Ba ($N = 88$), ^{146}Ba ($N = 90$) and ^{148}Ba ($N = 92$), $\beta_3 = 0.068$, 0.079 and 0.044, respectively.

2.4. Ground State Spins and Parities

Evidence of quadrupole-octupole shapes in a certain region of light actinide nuclei was found from the ground state spin and parity property of these nuclei [21, 22].

The experimental values of the ground state spins and parities (J^π) of odd-A (odd-N and even-Z and odd-Z and even-N) nuclei beyond ^{208}Pb have been determined, in a number of these nuclei. The spins are mostly obtained from measurements, like atomic beam laser spectroscopy. Other arguments for J^π assignments used are: favored α decay (low α hindrance factors) and favored beta-decay log ft values to levels with known J^π etc. For the details of J^π assignments for ground states in these nuclei refer to [17]. The experimental J^π values, in a certain region of actinide nuclei, in general, were found to be inconsistent with the predictions of the quadrupole Nilsson model. However, in the same nuclear region, the J^π predictions by model calculations [23, 24] with quadrupole-octupole nuclear shape were found to be consistent with the experimental values of ground state spins and parities. The details of calculations for J^π predictions can be found in the above two references, in which, in the former [23], the folded Yukawa potential and in the latter [24], the Woods–Saxon potential were used, for the quadrupole-octupole deformed Nilsson diagrams for neutrons and protons. The J^π predictions are similar to these two calculations when the quadrupole and octupole deformations are present. For further details see the excellent review article [22].

In Figs. 2.7 and 2.8 single neutron levels and single proton levels in a folded Yukawa potential with spheroidal symmetry ($\varepsilon_3 = 0$) are plotted as a function of quadrupole deformation ε_2 (left-hand side). On the right side, are plotted the parity mixed (neutron/proton) levels in an axially symmetric, reflection asymmetric, folded Yukawa potential ($\varepsilon_3 = 0.08$) as a function of quadrupole deformation ε_2.

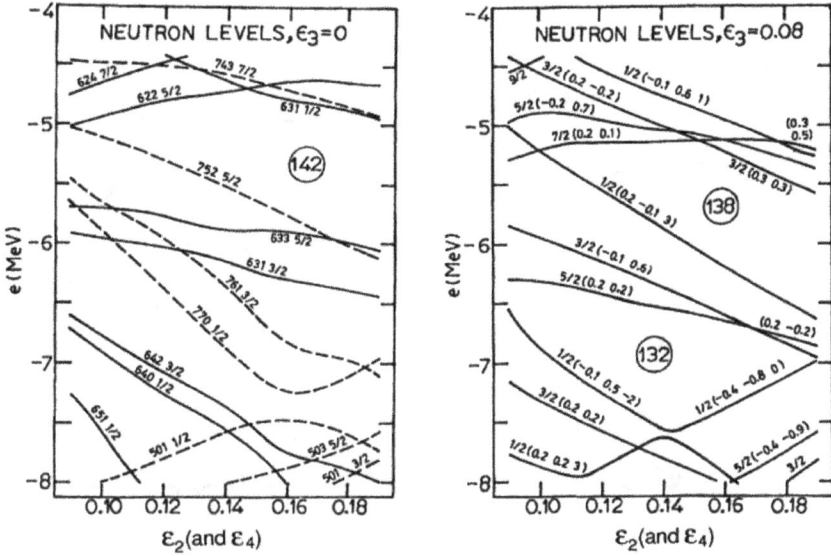

Fig. 2.7. (Left) Single neutron levels in the actinide region in a folded Yukawa potential with spheroidal symmetry ($\varepsilon_3 = 0$), plotted as a function of quadrupole deformation ε_2. Positive parity levels are represented by solid lines while the negative parity levels by dashed lines. These are labeled by the usual asymptotic quantum numbers $[Nn_z\Lambda]\Omega$. (Right) The parity mixed single neutron levels in the actinide region in an axially symmetric reflection asymmetric folded Yukawa potential ($\varepsilon_3 = 0.08$) are plotted as a function of quadrupole deformation ε_2. Here, the levels are labeled by Ω and in parentheses are a set of matrix elements: $(\langle \hat{s}_z \rangle, \langle \pi \rangle, \langle -j_+ \rangle \, \delta_{\Omega 1/2})$. For description of these matrix elements refer to [22]. The set of ε_4 values used is defined in [23]. Figure is reproduced with permission from [22].

The value $\varepsilon_3 = 0.08$ has been used as a mean value for octupole deformation in this region. The labeling of the levels are mentioned in the figure captions.

In Fig. 2.9, in the region of interest, a summary of the experimental values of spins and parities is given. The region of nuclei in which, in general, the ground-state J^π values are consistent with the quadrupole-octupole model and inconsistent with the quadrupole Nilsson model, is marked with solid and dashed lines as in [22]. For specific details of the quadrupole-octupole deformed neutron and proton orbital assignments for different Z and N values in this region,

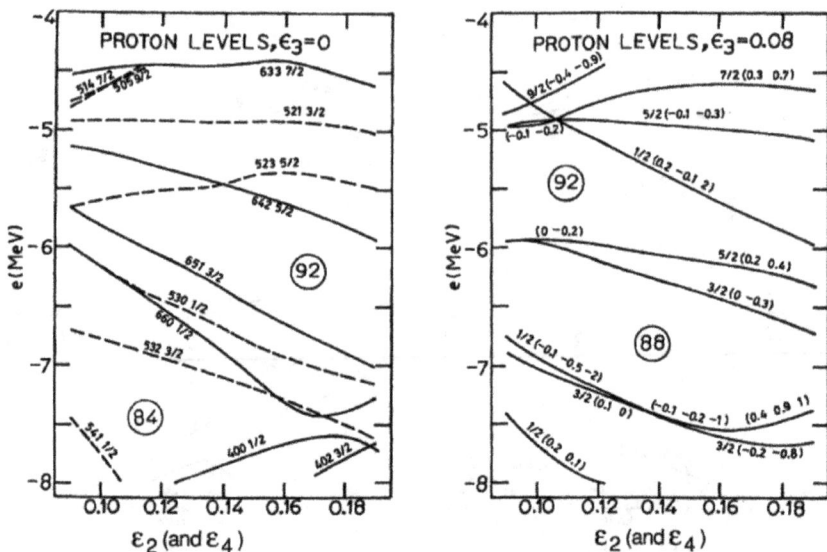

Fig. 2.8. (Left) Single proton levels in the actinide region in a folded Yukawa potential with spheroidal symmetry ($\varepsilon_3 = 0$), plotted as a function of quadrupole deformation ε_2. Positive parity levels are represented by solid lines while the negative parity levels by dashed lines. These are labeled by the usual asymptotic quantum numbers $[Nn_z\Lambda]\Omega$. (Right) The parity mixed single proton levels in the actinide region in an axially symmetric reflection asymmetric folded Yukawa potential ($\varepsilon_3 = 0.08$) are plotted as a function of quadrupole deformation ε_2. Here, the levels are labeled by Ω and in parentheses are a set of matrix elements: $(\langle \hat{s}_z \rangle, \langle \pi \rangle, \langle -\hat{j}_+ \rangle \delta_{\Omega 1/2})$. For description of these matrix elements refer to [22]. The set of ε_4 values used is defined in [23]. Figure is reproduced with permission from [22].

the reader is referred to the above reference. This marked region then defines the quadrupole-octupole deformed actinide nuclei.

In [25] evidence for ground state octupole deformation in the lanthanide region is discussed.

2.5. Nuclear Charge Radii

In atomic transitions, involving specific s electrons, the energy of the transition is dependent upon the neutron number N of the nucleus. The energy changes slightly between two isotopes. This isotope shift results because of the change in nuclear masses and change of the

N →	126	127	128	129	130	131	132	133	134	135	136	137	138	139	140	141	142	(proton orbital)
82Pb		9/2+		9/2+														
83Bi	9/2-		9/2-		9/2-													
84Po	9/2-	9/2+	9/2-	9/2+	9/2-	9/2+												
85At	9/2-		9/2-		9/2-		9/2-											
86Rn	(9/2+)	9/2+	9/2+	9/2+	9/2+	9/2+	5/2+	5/2+		7/2+		7/2+		7/2-				1/2 (-0.1, -0.5, -2) 3/2 (0.1, 0)
87Fr	9/2-	9/2+	9/2-		9/2-	(7/2)+	9/2- K=1/2		5/2- K=1/2		3/2(+)	1/2+	3/2-		1/2+		1/2+	
88Ra			9/2-		9/2	(7/2)+	(3/2)	5/2+		3/2	(3/2)	1/2+	3/2	3/2-	(3/2+)	5/2+	1/2+	
89Ac	9/2-	(9/2+)	9/2-		9/2		(3/2)	(5/2)+	(5/2)		(3/2)	3/2	3/2	(3/2-)	(3/2+)		1/2+	3/2 (0, -0.3)
90Th	9/2-	(9/2+)	9/2-		9/2	(7/2-)	(5/2)+	(5/2)+	(5/2)	(3/2-)	(3/2)	(1/2+)	5/2-	5/2-	(3/2+)	5/2+		
91Pa		(9/2+)	9/2-		9/2				(5/2)		(5/2)		(5/2+)	(5/2-)	3/2- K=1/2		3/2- K=1/2	5/2 (0.2, 0.4)
92U				(9/2+)						(3/2-)	(5/2)	(3/2-)	(5/2+)	(5/2)	(5/2+)	(5/2+)	5/2+	
93Np							(5/2)						(5/2)		(5/2+)		5/2+	1/2 (0.2, -0.1, 2)

Neutron orbitals (bottom): 3/2 (0.2, 0.2) [N=129]; 1/2 (-0.1, 0.5, -2) [N=131]; 5/2 (0.2, 0.2) [N=133]; 3/2 (-0.1, 0.6) [N=135]; 1/2 (0.2, -0.1, 3) [N=137]; 3/2 (0.3, 0.3) [N=139]; 5/2 (-0.2, 0.7) [N=141]

Fig. 2.9. Experimental ground-state spins and parities (J^π) for the odd-A nuclei beyond ^{208}Pb ($Z = 82$) to ^{235}Np ($Z = 93$) for $N = 126$ to 142. The relevant quadrupole-octupole deformed neutron and proton orbitals are mentioned at the bottom and to the right of the figure, respectively. The experimental data on ground-state J^π are taken from [17]. The region of quadrupole-octupole deformation as predicted from consistency of experimental ground-state J^π with quadrupole-octupole model predictions is depicted by a central frame of solid and dashed lines (see text for details). This figure is similar to Fig. 32 in [22].

nuclear Coulomb field experienced by electrons involved in the atomic transition [26]. The changes in the latter contribution is the result of change in the nuclear charge distribution between the two isotopes. The charge distribution calculations involve the mean square nuclear radius $\langle r^2 \rangle$. From the measured isotope shifts the changes in the mean square charge radii $\delta \langle r^2 \rangle$ are evaluated. Nuclear charge radius is one of the fundamental nuclear parameters, precise measurements of this has been done throughout the periodic table, using different experimental methods, like high-energy elastic electron scattering, muonic atom X-rays, optical isotope shifts and K_α X-rays isotope shifts [27 and references therein]. Optical isotope shifts using modern precision laser spectroscopy is mainly responsible for obtaining the changes in the mean square nuclear charge radii $\delta \langle r^2 \rangle$ in nuclei away from the line of beta-stability. The first two methods provide absolute root mean square nuclear charge radii $\langle r^2 \rangle^{1/2}$. N and Z dependence of nuclear charge radii in many of the isotopic chains has been discussed in detail in an excellent review article [28]. A table of nuclear ground state charge radii is available in [27]. Considering the systematics of nuclear charge radii in over 800 nuclei, an effective formula was suggested in [29] which takes into account the various observed effects. In [30] density functional theory has been used for a global description of the behavior of nuclear charge radii in nuclei to understand nuclear interactions.

The systematics of the isotopic variation of the mean square nuclear charge radii, in general, throughout the periodic table, exhibited a clearly visible remarkable feature, an odd-even staggering. The charge radii $\langle r^2 \rangle$ for the odd-N isotopes are found to be smaller ($\sim 10^{-2}$ fm^2) than the average of the radii of their even-even neighbors. This is the normal odd-even staggering. This may primarily be attributed to the pairing effects. The staggering can be discussed in terms of a staggering parameter Δ [31] as:

$$\Delta = \langle r^2 \rangle^N - (1/2)(\langle r^2 \rangle^{N+1} + \langle r^2 \rangle^{N-1}) \tag{2.1}$$

where the neutron number N is for the odd isotope. The above relation can also be expressed in $\delta \langle r^2 \rangle$ as:

$$\Delta = \delta \langle r^2 \rangle^{N-1,\, N} - (1/2)\delta \langle r^2 \rangle^{N-1,\, N+1} \tag{2.2}$$

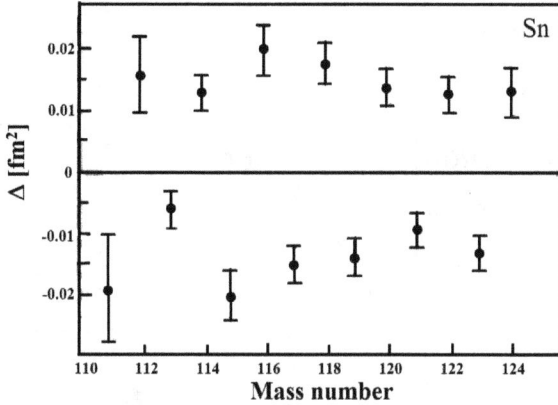

Fig. 2.10. Normal odd-even staggering in Sn isotopes (from $N = 61$ to 74) as a function of mass number A. The y-axis represents the odd-even staggering parameter Δ (see text). Figure adopted from [32].

An example of normal odd-even staggering in tin isotopes is shown in Fig. 2.10 [32], which is a plot of the staggering parameter Δ as a function of mass number A for $N = 61$ to 74. This clearly depicts the staggering of the mean square charge radii between the odd and the even tin isotopes.

The mean square charge radius is a function of nuclear charge distribution which is dependent upon nuclear shell closure (spherical nuclear shape) or nuclear deformation (quadrupole or quadrupole-octupole shape). The following expression for deformation dependence of change in mean square charge radii can be written [31, 33]:

$$\delta\langle r^2 \rangle = \delta\langle r^2 \rangle_0 + \frac{5}{4\pi}\langle r^2 \rangle_0 \, \Sigma_\lambda \, \delta\langle \beta_\lambda^2 \rangle \tag{2.3}$$

where $\langle r^2 \rangle_0$ is the spherical droplet model radius and $\delta\langle \beta_\lambda^2 \rangle$ the change in the mean square deformation of multipolarity $\lambda(\lambda = 2, 3, 4, \ldots)$.

The normal observed odd-even staggering thus can be disturbed by deformation, in magnitude and sign. Experimental investigations of changes in nuclear charge radii $\delta\langle r^2 \rangle$ have been carried out in the Pb region as a function of neutron number [34 and references therein]. For the lighter Po isotopes with $N < 115$, deviations from the trend for the heavier isotopes was found. This was related to changes

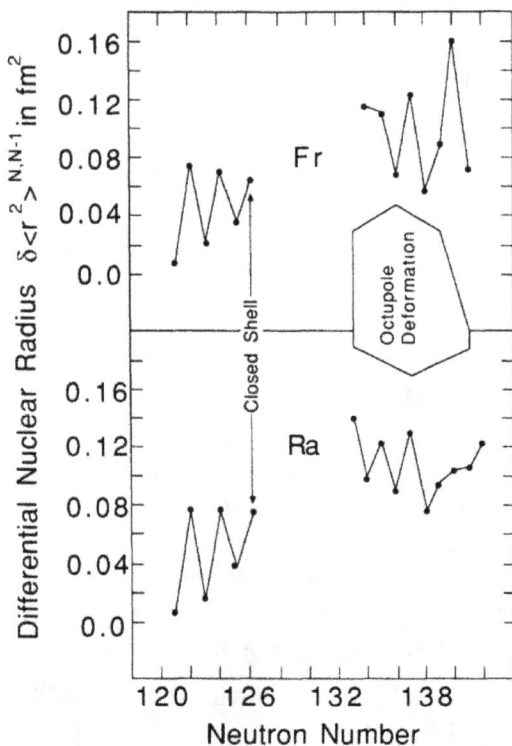

Fig. 2.11. Plot of odd-even staggering in the differential nuclear charge radii $\delta\langle r^2\rangle^{N,\,N-1}$ as a function of neutron number N, in the Fr and Ra isotopes, normal odd-even staggering is found below and inverted odd-even staggering above the neutron magic number $N = 126$. Notice the gap in data points between $N = 126$ to 132 as the isotopes in this range cannot be online mass separated with present techniques because of short (ms to μs) halflives of the ground states of these isotopes. $\delta\langle r^2\rangle^{N,\,N-1}$ is the difference in the mean square charge radius $\langle r^2\rangle$ for a nucleus with neutron number N and with $N-1$. This figure has been reproduced with permission from [35].

in quadrupole deformation. In the light actinide region of nuclei where there is experimental evidence from nuclear spectroscopic measurements for the nuclei to be octupole deformed, the odd-even staggering is found to be inverted in sign. This is called inverted odd-even staggering. It is depicted in Fig. 2.11 [35] for the Fr and Ra isotopes. See also [36] for recent laser spectroscopic measurements for changes in mean-square charge radii in some Ra isotopes. A

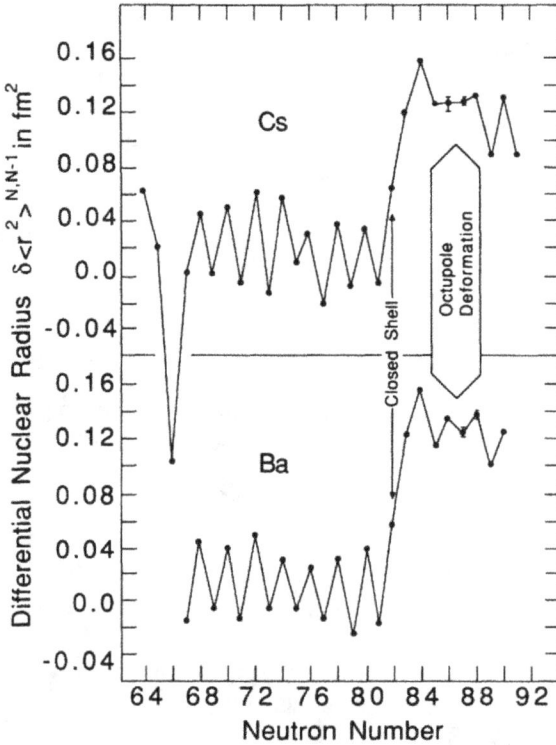

Fig. 2.12. Plots of odd-even staggering of differential nuclear charge radii $\delta\langle r^2 \rangle^{N,\,N-1}$ as a function of neutron number for the odd and even isotopes of Cs and Ba nuclei. See text for details. The figure is reproduced with permission from [35].

discussion on the observation of normal and inverted (inverse) odd-even staggering for Fr, Ra and Pb, Po, At, Fr, Rn and Ra nuclei is presented in [36] and [37] respectively.

In Fig. 2.12 are shown plots of differential charge radii $\delta\langle r^2 \rangle^{N,\,N-1}$ as a function of neutron number N for the isotopic chains in Cs and the Ba nuclei [35]. Normal odd-even staggering is seen in both the nuclei below the neutron magic number $N = 82$. A large decrease in the differential charge radius occurs at $N = 66$ because of sudden appreciable decrease in quadrupole deformation due to structural changes. After neutron number $N = 82$, between $N = 85 - 88$, the normal odd-even staggering is absent in Cs isotopes and is attenuated in the Ba isotopes. This effect though is not as spectacular (in inverted odd-even staggering) as seen in the actinides, has been attributed to octupole deformation.

References

1. P.A. Butler and W. Nazarewicz, *Rev. Mod. Phys.* **68**, 349 (1996) and references therein.
2. P.A. Butler, *J. Phys. G: Nucl. Part. Phys.* **43**, 073002 (2016) and references therein.
3. S.E. Agbemava, A.V. Afanasjev and P. Ring, *Phys. Rev. C* **93**, 044304 (2016) and references therein.
4. S.E. Agbemava and A.V. Afanasjev, *Phys. Rev. C* **96**, 024301 (2017).
5. A.V. Afanasjev, H. Abusara and S.E. Agbemava, *Phys. Scr.* **93**, 034002 (2018).
6. K. Nomura *et al.*, *Phys. Rev. C* **97**, 024317 (2018).
7. S.Y. Xia *et al.*, *Phys. Rev. C* **96**, 054303 (2017).
8. Z.P. Li *et al.*, *J. Phys. G: Nucl. Part. Phys.* **43**, 024005 (2016).
9. J.M. Yao *et al.*, *Phys. Rev. C* **92**, 041304 (R) (2015).
10. H.-L. Wang *et al.*, *Phys. Rev. C* **92**, 024303 (2015).
11. Z.P. Li *et al.*, *Phys. Lett. B* **726**, 866 (2013).
12. J.Y. Guo *et al.*, *Phys. Rev. C* **82**, 047301 (2010).
13. Y. Fu *et al.*, *Phys. Rev. C* **97**, 024338 (2018).
14. W. Zhang *et al.*, *Chin. Phys. C* **34**, 1094 (2010).
15. K. Nomura *et al.*, *Phys. Rev. C* **89**, 024312 (2014).
16. K. Nomura *et al.*, *Phys. Rev C* **88**, 021303 (R) (2013).
17. Brookhaven National Nuclear Data Center, ENSDF files: http://www.nndc.bnl.gov and references therein.
18. priv. comm. from Professor A.V. Afanasjev, June 2018.
19. W. Nazarewicz *et al.*, *Nucl. Phys. A* **429**, 269 (1984).
20. S. Raman *et al.*, *At. Data Nucl. Data Tables* **78, 1** (2001).
21. R.K. Sheline, *Phys. Lett. B* **197**, 500 (1987).
22. A.K. Jain *et al.*, *Rev. Mod. Phys.* **62**, 393 (1990).
23. G.A. Leander and R.K. Sheline, *Nucl. Phys. A* **413**, 375 (1984).
24. G.A. Leander and Y.S. Chen, *Phys. Rev. C* **37**, 2744 (1988).
25. G.A. Leander *et al.*, *Phys. Lett. B* **152**, 284 (1985).
26. K. Heilig and A. Steudel, *At. Data and Nucl. Data Tables* **14**, 613 (1974).
27. I. Angeli and K.P. Marinova, *At. Data and Nucl. Data Tables* **99**, 69 (2013) and references therein.
28. I. Angeli *et al.*, *J. Phys. G* **36**, 085102 (2009).
29. Z. Sheng *et al.*, *Eur. Phys. J. A* **51**, 40 (2015).
30. P.-G. Reinhard and W. Nazarewicz, *Phys. Rev. C* **95**, 064328 (2017).
31. W. Kälber *et al.*, *Z. Phys. A* **334**, 103 (1989).
32. M. Anselment *et al.*, *Phys. Rev. C* **34**, 1052 (1986).

33. S.A. Ahmad *et al.*, *Nucl. Phys. A* **483**, 244 (1988).
34. M.D. Seliverstov *et al.*, *Phys. Lett. B* **719**, 362 (2013) and references therein.
35. R.K. Sheline, A.K. Jain and K. Jain, *Phys. Rev. C* **38**, 2952 (1988).
36. K.M. Lynch *et al.*, *Phys. Rev. C* **97**, 024309 (2018).
37. A.E. Barzakh *et al.*, *Phys. Rev C* **99**, 054317 (2019).

Chapter 3

High Spin Behavior of Pear-Shaped Nuclei-I

3.1. Introduction

In this chapter, we will try to gather information about the behavior and properties of pear-shaped nuclei at high spins, from the experimentally measured excitation energy of the negative parity and the positive parity levels populated in some of the actinide and the lanthanide nuclei, as a function of spin and/or rotational frequency. In Sec. 3.2, a discussion is given on the excitation energy systematics of levels in the negative parity bands in relation to the positive parity levels in the ground state bands in a number of even-even isotopes in Ra, Th, Ba and Sm nuclei. Section 3.3 is devoted to the excitation energy splittings between the negative and the positive parity bands at low spins and the formation of alternating parity bands after a certain spin value, in these nuclei. In Sec. 3.4, attention is paid to the displacement energies or the parity splittings in these nuclei to know about the stabilization of octupole shapes as a function of spin. It is interesting to know from the nuclei exhibiting octupole correlations, which are nuclei that are octupole vibrational and which are octupole deformed. For this purpose, in Sec. 3.5.1, the ratio of rotational frequencies ω_{rot} $(\pi = -)/\omega_{\text{rot}}$ $(\pi = +)$ for the negative and the positive parity bands as a function of spin in the even-even nuclei, is considered. In Sec. 3.5.2, we take into consideration $E(J^\pi)/E(2^+)$ energy ratios versus spin I for the states in the positive parity ground state and negative parity octupole bands. In Sec. 3.5.3, aligned angular momentum i and the difference in the aligned angular momentum, Δi_x, for the negative and the positive parity bands, are discussed. To get an insight into the rotational behavior of the octupole nuclei, in Sec. 3.6, the total aligned angular momentum and the kinematic moment of inertia as a function of rotational frequency for the negative and the positive parity bands, are considered.

In an odd-N or odd-Z nucleus, the motion of the odd particle (neutron or proton) and its coupling to the quadrupole-octupole deformed even-even core determines the observed nuclear structure. As an analogy from molecular spectroscopy, in such systems, the combination of the intrinsic parity of the particle and that of the axially symmetric reflection asymmetric deformed core gives rise to nearly degenerate pairs of states of opposite parity — the parity doublets. In Sec. 3.7, a discussion on these parity doublet bands observed in actinide nuclei is presented. The evolution of octupole correlations in even-even nuclei has been considered in Sec. 3.4 through the behavior of displacement energy or parity-splitting as a function of spin. The same aspect has been looked into in some of the odd-N actinide nuclei in Sec. 3.8. In the following section (Sec. 3.9) attention has been paid to the behavior of kinematic moment of inertia as a function of rotational frequency. The last Sec. 3.10 deals with the low energy predominantly magnetic dipole ($M1$) transitions observed in ^{223}Ra and ^{223}Th nuclei and the associated magnetic dipole moments of the relevant excited states.

A. Even-Even Nuclei

3.2. Excitation Energy Systematics of Levels

In this section, excitation energy systematics of negative parity octupole band levels in relation to the ground state positive parity band levels as a function of neutron number, will be considered. For this purpose, even-even Ra ($Z = 88$), Th ($Z = 90$) and the Ba ($Z = 56$) and Sm ($Z = 62$) nuclei have been chosen as representative of $A \sim 220$ and $A \sim 150$ mass regions.

In Fig. 3.1 are shown the excitation energies of the even-spin positive parity levels 2^+, 4^+, 6^+, 8^+ and 10^+ of the ground state ($K^\pi = 0^+$) bands and the odd-spin negative parity levels 1^-, 3^-, 5^-, 7^-, and 9^- of the octupole ($K^\pi = 0^-$) bands, as a function of neutron number N, for the even-even Ra and Th isotopes from $N = 128$ to 142. As we move beyond the neutron magic number or closed neutron shell $N = 126$, the excitation energies of the positive parity states 2^+, 4^+, ... decrease in a regular manner. This is similar to the well-known phenomenon that was observed for the quadrupole deformed even-even rare earth nuclei in the mass region $A \sim 160$ where it is due to the evolution or increase in quadrupole collectivity with increase in neutron number, signifying a transition of the nuclear shape to stable quadrupole deformation. It is the manifestation of the same effect on these light actinide even-even nuclei. What is interesting is that not only is there an increase in quadrupolar collectivity, for these lighter Ra and Th isotopes, but simultaneously in parallel, the octupolar collectivity also increases as shown by a parallel decrease of excitation energies of the negative parity states till $N \sim 134$. This should also then indicate a transition towards stable octupole deformation. The excitation energy plot for the negative parity states, exhibit a near parabolic shape as a function of neutron number. For Ra isotopes, the minima for the shape occur in the region $N = 134$ to 138 for the 1^-, 3^- and the 5^- states, that is, the excitation energies are lowest for these states at these neutron numbers. In the case of Th isotopes, the minimum is a bit sharper (see lower panel in the figure), the energies of the negative parity 1^-, 3^- and the 5^- states show the lowest values between $N \sim 134$ to 136. Comparing the excitation energies of the 1^- states in the Ra and Th isotopes, in ^{224}Ra $(N = 136)$ 1^- state has the lowest energy 215.98 keV and in ^{226}Th $(N = 136)$ nucleus, it is 230.37 keV. Beyond neutron numbers $N = 138$ in Ra and $N = 136$ in Th, and before $N = 134$ in both nuclei, the excitation energies of the negative parity states steeply increase indicating a decrease in octupole collectivity. It may further be noted that in ^{218}Ra, the 3^- state is lowest in energy and not the 1^- state.

Similar to Fig. 3.1, excitation energies are plotted in Fig. 3.2, of the even-spin positive parity levels 2^+, 4^+, 6^+, 8^+ and 10^+ of the ground state bands and the odd-spin negative parity levels 1^-, 3^-, 5^-, 7^-, and 9^- of the octupole bands, as a function of neutron number N, for the even-even Ba and Sm isotopes from $N=84$ to 92 and 84 to 94 respectively. Here, the variation of excitation energies

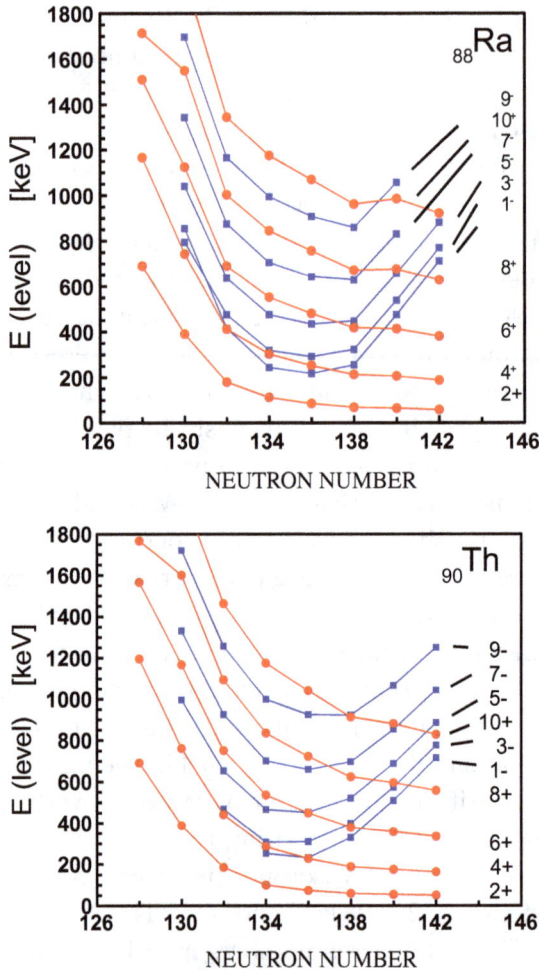

Fig. 3.1. Energy level systematics of the positive parity ground state bands and the negative parity octupole bands in the even-even Ra (upper panel) and Th (lower panel) $N = 128$–142 isotopes as a function of neutron number. The data are taken from [13 and references therein]. The energy level for the 1^- state in ^{218}Ra is from [14].

with neutron number is not that spectacular as in the case of Ra and Th isotopes. In Ba isotopes (upper panel), the excitation energies do decrease sharply up to $N = 88$. This decrease is similar for the positive and the negative parity band levels, signifying the

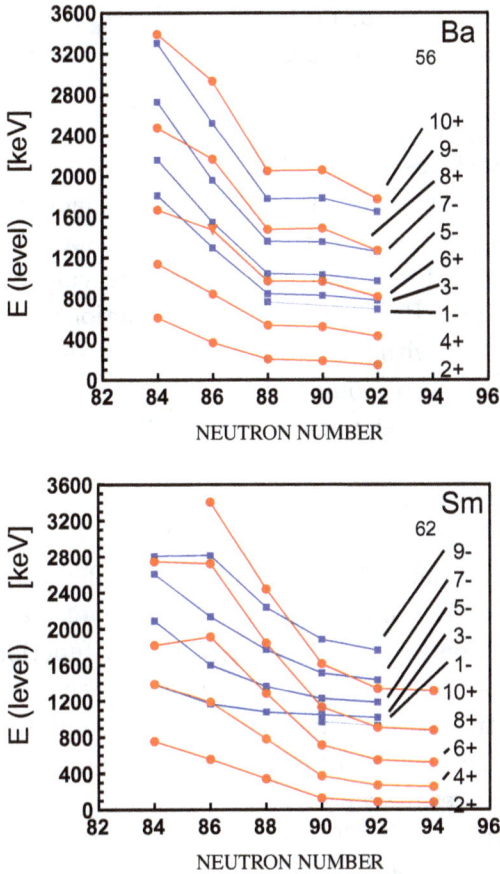

Fig. 3.2. Energy level systematics of the positive parity ground state bands and the negative parity octupole bands in the even-even Ba (upper panel) $N = 84$–92 and Sm (lower panel) $N = 84$–94 isotopes as a function of neutron number. The data are taken from [13 and references therein].

evolution of both the quadrupolar and the octupolar collectivity in the same neutron region. Here, for the negative parity states, there is no parabolic shape for the variation of the excitation energies with neutron number as in the case of Ra and Th nuclei. In the $N = 88$ to 92 region, for the Ba isotopes, the 1^- to 9^- states are lowest in energy with a small variation. For the Sm isotopes (lower panel), the situation is better, there is a sharp decrease of the excitation energies for both the positive and the negative parity states signifying the simultaneous evolution of the quadrupolar and octupolar collectivities. Between $N = 90$ to 94, the variation of excitation energies is slow. There is no parabolic shape for the negative parity states till $N = 94$.

A number of attempts [1–12] have been made to explain these excitation energy variations with neutron number. In [9], the energies of 1^- states for the Ra isotopes were calculated using the Barcelona–Catania–Paris energy density functionals and in the same theoretical manner with the Gogny DIS interaction. The theoretically calculated values were compared with the experimental data (see Fig. 7 in the above reference). In Fig. 2 of ref. [10], the authors have compared their results of theoretical calculations using two-dimensional Q20–Q30 generator coordinate method (GCM), one-dimensional Q3 GCM and the Gogny DIS (DIM) interactions, with the experimental data also for the 1^- states for the even-even Ra and Th isotopes. Recently, detailed calculations have been done [11] which are based on microscopic energy density functional framework for the excitation energies of the even-spin positive parity and the odd-spin negative parity band states in the light Ra, Th and Ba, Sm even-even isotopes as a function of mass number, and compared their theoretical results with the experimental data (Figs. 11 and 12 in the above reference). The evolution of excitation energies of the positive parity states of the ground state band and the negative parity octupole band states and their values are in agreement with the data for the lower spin states. An in-depth discussion of the results of the calculations is given in this work [11].

3.3. Energy Splittings and Alternating Parity Bands

One of the most characteristic features or signature of reflection-asymmetric shape in an even-even nucleus is the observation of close-lying even-spin positive parity $J^\pi = 0^+, 2^+, 4^+, \ldots$ band and an odd-spin negative parity $J^\pi = 1^-, 3^-, 5^-, \ldots$ band, in specific mass regions. The alternating parity states, for example, like the 5^- and 4^+ states or 6^+ and 5^- states, are connected by strong $\Delta J = 1$,

$E1$ transitions. These $E1$ transitions compete with the in-band $E2$ transitions. In general, for the lighter mass isotopes in these nuclei, above $J \sim 7\,\hbar$ (in light actinides), the positive and negative parity states approximately form interleaved alternating parity $J = 0^+$, 1^-, 2^+, 3^-, 4^+, 5^-, ... band. In the following, this situation is considered in detail as a function of angular momentum in even-even actinide and the lanthanide nuclei.

Figures 3.3–3.8 are plots of yrast lines formed by the excitation energies of the even-spin positive parity ground state band states and the odd-spin negative parity band states in the even-even Ra, Th, Ba and Sm isotopes. In Figs. 3.3 and 3.4, the plots are for the even-even Ra ($Z = 88$) isotopes for ^{218}Ra ($N = 130$) to ^{228}Ra ($N = 140$). It is observed that at low spins, the positive parity even spin yrast line is pushed down when compared to the negative parity odd spin yrast line. In $N = 130$ to 138 nuclei, at $J = 3$, this energy difference varies between 110–230 keV and in ^{228}Ra ($N = 140$) it is $\sim$400 keV. This energy difference decreases gradually as the angular momentum increases. It nearly vanishes at $J \sim 6$ to $9\,\hbar$ in $N = 130$, 132, 134 and 136 in these Ra nuclei. The situation is different in the heavier isotopes, in ^{226}Ra ($N = 138$) and ^{228}Ra ($N = 140$), the energy difference tends to disappear only at $J \sim 11\,\hbar$ and $18\,\hbar$, respectively.

The even-even Ra isotopes, ^{218}Ra ($N = 130$) and ^{220}Ra ($N = 132$) have been investigated [15, 16] to very high spins $J \sim 30\,\hbar$. In ^{218}Ra, the quadrupole deformation β is $\sim$0.095 as determined from the measured $B(E2; 0^+ \rightarrow 2^+)$ [17 and references therein]. After $J \sim 7\,\hbar$ in this nucleus, there is a reversal of the yrast lines, the negative parity yrast line now lies lower than the positive parity yrast line, a sign inversion. The energy splitting here is much smaller as compared to that at low spin. This inversion continues to very high spins, at $J \sim 29\,\hbar$, there is then an indication of another sign inversion. In ^{220}Ra, however, no measured value of β deformation is available, the experimental $E(4^+)/E(2^+)$ ratio is 2.29 as compared to 1.90 in ^{218}Ra. This indicates that the quadrupole deformation in ^{220}Ra is higher near the ground state than in ^{218}Ra. Also, in ^{220}Ra, well deformed positive parity ground state and negative parity octupole deformed bands have been found [16]. A close look of the

Pear-Shaped Nuclei

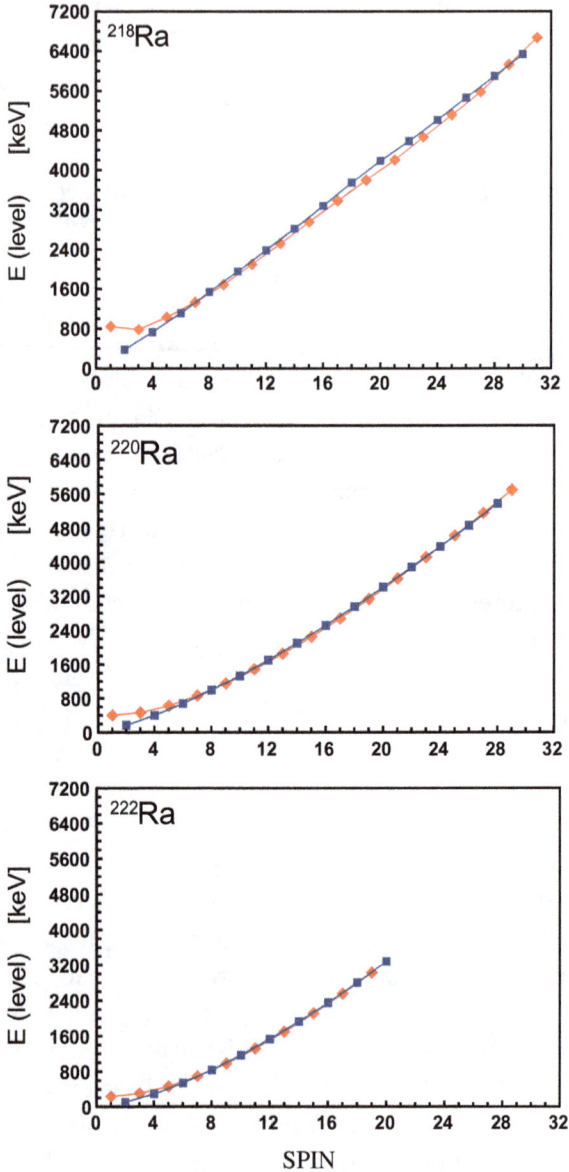

Fig. 3.3. Plot of excitation energy of the positive parity ground state even-spin band levels (■) and the negative parity odd-spin octupole band levels (♦) as a function of spin in even-even $N = 130$, 132 and 134 Ra ($Z = 88$) isotopes. Experimental data are taken from [13 and references therein]. The energy level for the 1^- state in ^{218}Ra is from [14].

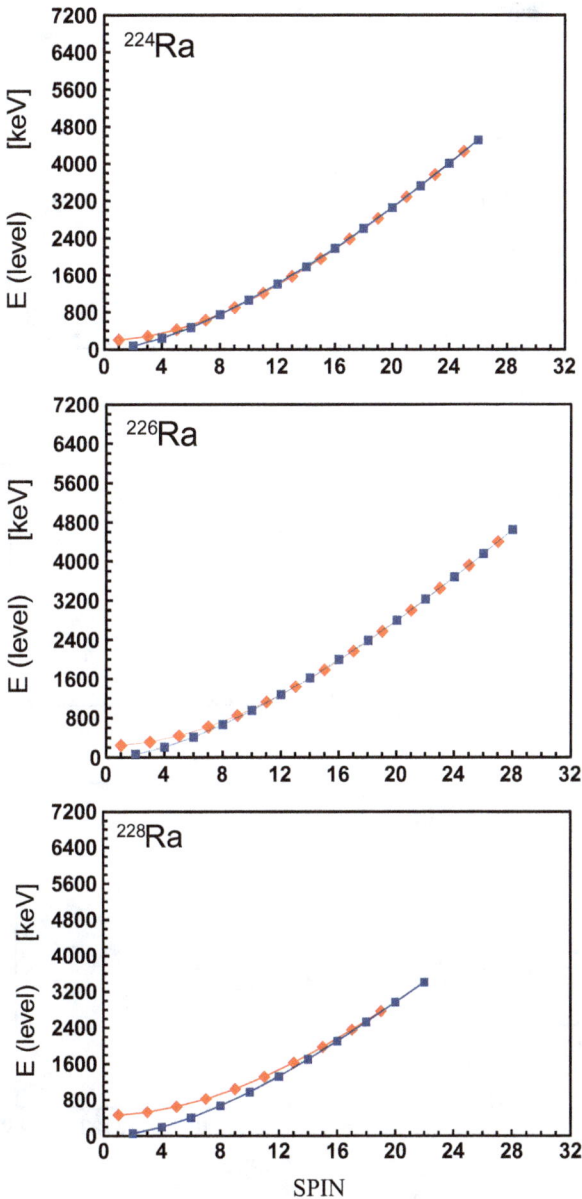

Fig. 3.4. Same as for Fig. 3.3, but for $N = 136$, 138 and 140 even-even Ra isotopes. Experimental data for ^{224}Ra and ^{226}Ra are taken from [19] and for ^{228}Ra from [13 and references therein].

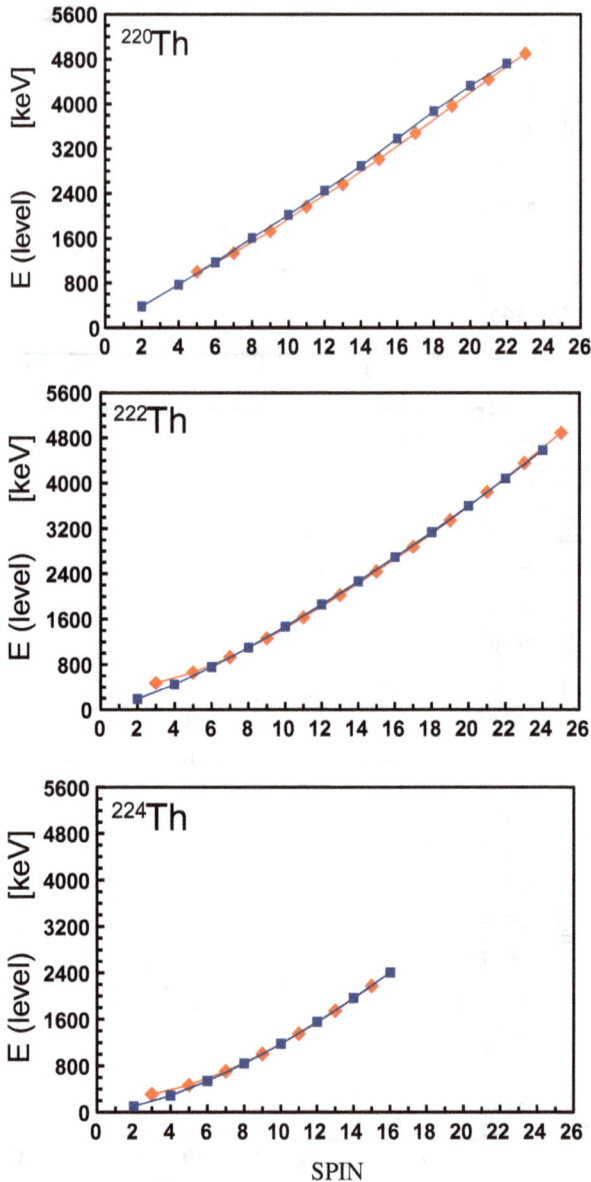

Fig. 3.5. Plot of excitation energy of the positive parity even-spin ground state band levels (■) and the negative parity odd-spin octupole band levels (♦) as a function of spin in even-even $N = 130$, 132 and 134 Th ($Z = 90$) isotopes. Experimental data are taken from [13].

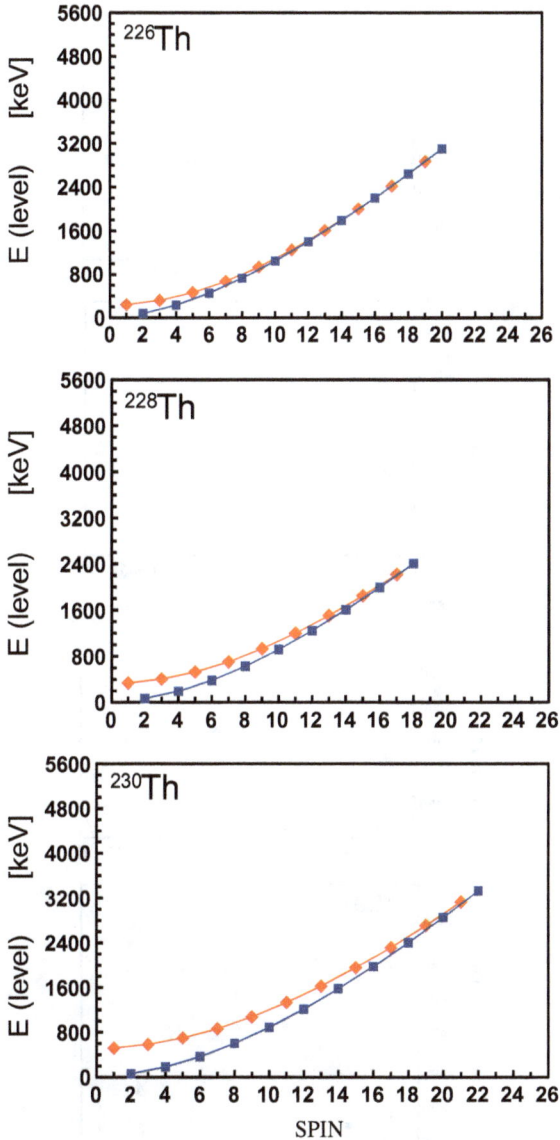

Fig. 3.6. Same as for Fig. 3.5, but for $N = 136$, 138 and 140 even-even Th isotopes. Experimental data are taken from [13].

Pear-Shaped Nuclei

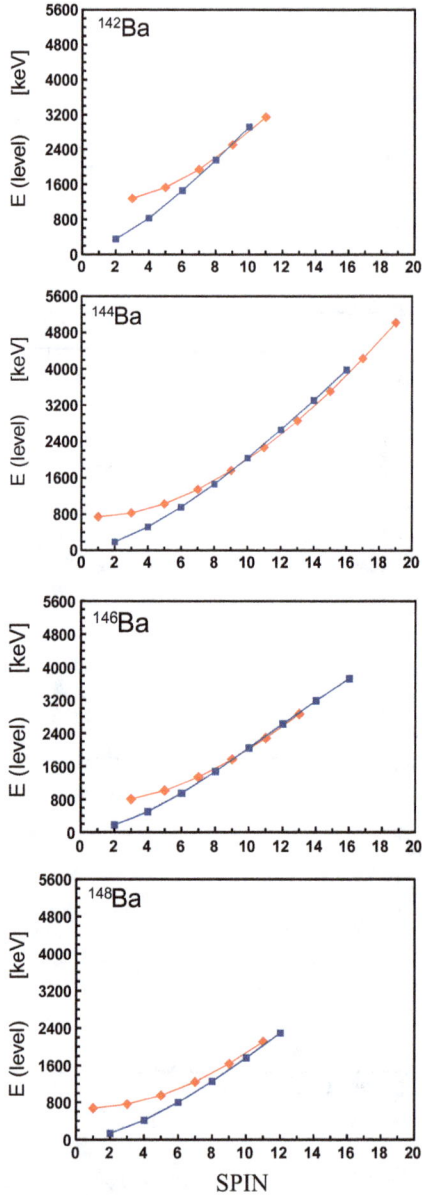

Fig. 3.7. Plot of excitation energy of the positive parity even-spin ground state band levels (■) and the negative parity odd-spin octupole band levels (♦) as a function of spin in even-even N = 86, 88, 90 and 92 Ba (Z = 56) isotopes. Experimental data are taken from [13].

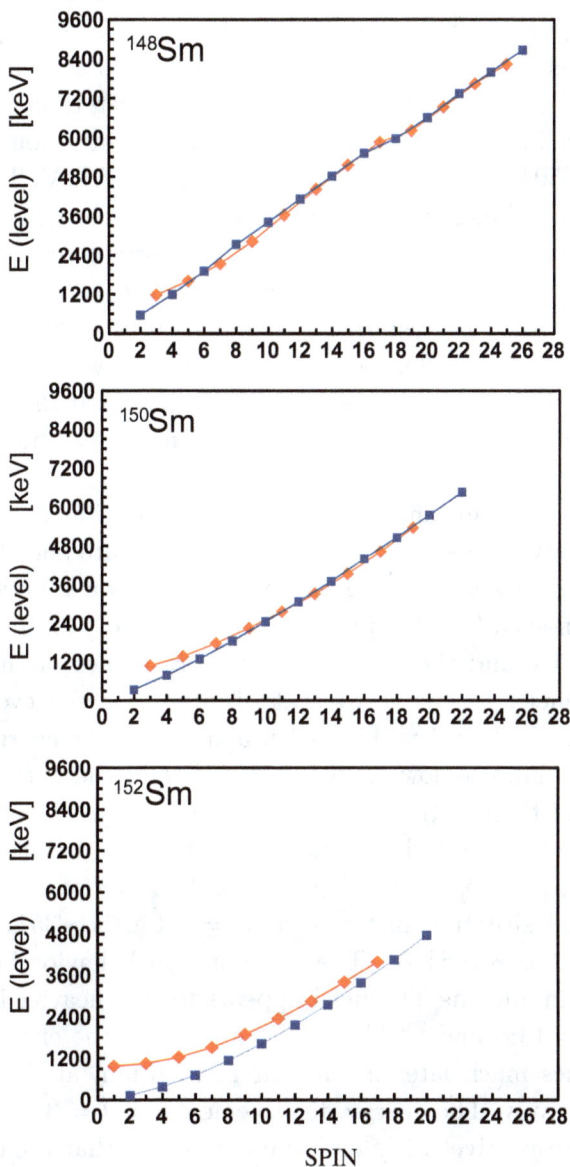

Fig. 3.8. Plot of excitation energy of the positive parity even-spin ground state band levels (■) and the negative parity odd spin octupole band levels (♦) as a function of spin in even-even $N = 86$, 88 and 90 Sm ($Z = 62$) isotopes. Experimental data are taken from [13].

yrast lines plot in ^{220}Ra (Fig. 3.3 middle panel) in an expended xy-view, reveals that after $J \sim 9\,\hbar$, the yrast lines are inversed, that is, negative parity band yrast line is then lower than the positive parity yrast line. The energy separation in this situation is small, at $J = 16\,\hbar \sim 50\,\text{keV}$ whereas, at $J = 3\,\hbar$, it is $\sim 180\,\text{keV}$. The inversed separation continues up to $J \sim 24\,\hbar$, it then changes sign again. In the reflection asymmetric shell model formulated in Ref. [18], the authors, using their model, calculated the energy of states in the positive and the negative parity bands as a function of spin in the even-even Ra isotopes (see their Fig. 2 which is similar to Figs. 3.3 and 3.4 in this sub-section) and compared the results with experimental data. The agreement between theory and experiment is excellent.

In Figs. 3.5 and 3.6, are plotted the excitation energies of the positive parity even-spin states of the ground state and the negative parity odd-spin states of the octupole band, as a function of spin, for the even-even Th isotopes from $N = 130$ to 140. The behavior of the positive and the negative parity yrast lines of the bands in these Th nuclei is similar to that observed for the even-even Ra isotopes. In ^{220}Th and ^{222}Th nuclei, apart from the energy splitting of the yrast lines at low spins, at $J \sim 6\text{--}8\,\hbar$, the energy splitting vanishes and there is the sign inversion of the yrast lines up to high spins ($J \sim 22\,\hbar$). Only these two (^{220}Th and ^{222}Th) nuclei have been investigated [16, 20] to very high spins. The energy splitting in the sign inverted situation in ^{222}Th at $J = 14\,\hbar$, is $\sim 35\,\text{keV}$ whereas at $J = 3$, it is $\sim 155\,\text{keV}$. The difference in behavior between the even-even Ra and the Th nuclei appears for the heavier isotopes, in ^{228}Th ($N = 138$) and ^{230}Th ($N = 140$) nuclei, the energy splitting here vanishes much later in angular momentum, at $J \sim 17\,\hbar$ and $22\,\hbar$ respectively. The energy displacements at $J = 3$ are ~ 270 and $\sim 460\,\text{keV}$, respectively. It should be noted here that the quadrupole deformation β as determined from the experimental measurements of $B(E2; 0^+ \rightarrow 2^+)$ [17 and references therein] in these nuclei are higher ($\beta = 0.230$ in ^{228}Th and 0.244 in ^{230}Th) in comparison to the corresponding Ra isotones ($\beta = 0.202$ in ^{226}Ra and 0.217 in ^{228}Ra).

In ^{222}Ra ($N = 134$) and ^{226}Ra ($N = 138$) nuclei, in the sign inversed region, the energy separation is $\sim 16\,\text{keV}$ at $J = 12\,\hbar$ and

$J = 15\,\hbar$, respectively. In ^{224}Th ($N = 134$) nucleus, the energy separation in the sign inversion region at $J = 12\,\hbar$ is $\sim$7 keV. It is lowest $\sim$3 keV at $J = 17\,\hbar$ in ^{226}Th ($N = 136$) but there is no sign inversion of the positive and the negative parity yrast lines. These observations are in agreement with Fig. 3.1, where the energies of the even-spin positive parity and the odd-spin negative parity states are plotted as a function of neutron number for the Ra and Th even-even nuclei. Here, the negative parity states are lowest in energy for $N = 134$ and $N = 136$ Ra and Th nuclei and near lowest in ^{226}Ra ($N = 138$) nucleus, indicating that the octupole correlations are strongest in these nuclei.

Let us now look at the behavior of the yrast lines for the ground state and the octupole bands in $A \sim 150$ even-even nuclei. In Fig. 3.7 are plotted the excitation energies of the positive parity even-spin states of the ground state and the negative parity odd-spin states of the octupole band in ^{142}Ba ($N = 86$) to ^{148}Ba ($N = 92$) nuclei as a function of spin. As for the actinide even-even nuclei considered above, the positive parity yrast line is pushed down in energy at low spins when compared to the negative parity yrast line. The energy separation being larger, at $J = 3$, it is between $\sim$700 to 500 keV in these even-even Ba nuclei. The energy separation between the yrast lines vanishes at $J \sim 9$–$10\,\hbar$ in the $N = 86$, 88 and 90 nuclei. ^{144}Ba has been investigated in [21, 22]. In this nucleus, above this spin value, there is sign inversion of the yrast lines, the negative parity band yrast line goes below the positive parity band yrast line. The energy separation between the yrast lines after sign inversion is $\sim$130 keV at $J = 14\,\hbar$. The quadrupole deformation β as determined from the measured $B(E2; 0^+ \rightarrow 2^+)$ [17 and references therein] for ^{144}Ba is 0.194 and for ^{146}Ba, it is 0.218. In ^{148}Ba ($N = 92$) nucleus, the energy separation becomes small only near $J = 11\,\hbar$.

Figure 3.8 gives a plot of excitation energy of the positive parity states of the ground state band and the negative parity states of the octupole band in ^{148}Sm ($N = 86$), ^{150}Sm ($N = 88$) and ^{152}Sm ($N = 90$) even-even nuclei, as a function of spin. Similar to the nuclei discussed above, in these Sm nuclei, at low spins, the positive parity yrast line is pushed down when compared to the negative parity yrast line. In ^{150}Sm ($N = 88$), there is also a gradual decrease of energy separation of the yrast lines with increase in angular momentum. It vanishes at $J \sim 11\,\hbar$. After this spin value, there is sign inversion, the negative parity yrast line becomes lower than the positive parity yrast line. This continues to $J \sim 18\,\hbar$. At low spins, the energy separation is $\sim 530\,\text{keV}$ at $J = 3\,\hbar$ and after the sign inversion, at $J = 16\,\hbar$, it is $\sim 125\,\text{keV}$. In ^{152}Sm, at low spin ($J = 3\,\hbar$), the energy separation is about 800 keV, although it decreases with increase in spin but it does not vanish even until $J \sim 18\,\hbar$. The quadrupole deformation β as determined from the measured $B(E2; 0^+ \to 2^+)$ [17 and references therein] for ^{150}Sm, is 0.193 and for ^{152}Sm, it is 0.306.

3.4. Displacement Energy or Parity Splitting

To understand octupole correlations in nuclei, the energy displacement between the negative parity and the positive parity bands, was defined in [23]. The positive parity band has even spin levels with $J^\pi = 0^+$, 2^+, 4^+,... and the negative parity band has odd spin levels $J^\pi = 1^-$, 3^-, 5^-,... The energy difference between a state at a particular odd spin I in the negative parity band and a state at the same odd spin I in the positive parity band can be estimated in a simple manner by subtracting from the energy of negative parity odd spin state I, a value calculated by taking an average of the energies of the adjacent even spin states in the positive parity band. Since the two bands have opposite parity, the displacement energy is also known as parity splitting. The displacement energy

$$\delta E(I) = E(I^-) - [E\{(I+1)^+\} + E\{(I-1)^+\}]/2 \qquad (3.1)$$

Instead of taking an average value, one can take an interpolated value from the positive parity band

$$\delta E(I) = E(I^-) - E_{\text{int}}(I^+) \qquad (3.2)$$

In [24 and references therein], the interpolated value $E_{\text{int}}(I^+)$ is obtained by considering the $I(I+1)$ behavior for the positive parity band.

$$\delta E(I) = E(I^-) - [(I+1) \times \{E(I-1)^+\} + I \times \{E(I+1)^+\}]/(2I+1) \qquad (3.3)$$

In [25], the following equivalent expression for $\delta E(I)$ has been used:

$$\delta E(I) = E_\gamma(I^- \to (I-1)^+) - I \times [E_\gamma((I+1)^+ \to (I-1)^+)]/(2I+1)$$
$$(3.4)$$

In literature [26–29], some extended formulae for $\delta E(I)$ are also given. In the discussion below, we have used Eq. (3.4) for the calculation of displacement energy $\delta E(I)$.

The displacement energy $\delta E(I)$ as a function of spin I, for even-even Ra, Th, Ba, Sm nuclei and the $N = 88$ isotones, are shown in Figs. 3.9–3.11. In all these plots, there is one observation which appears to be a common general feature and it is that at low spins, the displacement energies decrease or approach the line with $\delta E(I) = 0$, as a function of increase of spin (or say rotational frequency). This zero value of displacement energy is expected for stable octupole deformation. Let us now look at the individual plots. For the even-even ^{218}Ra, ^{220}Ra, ^{222}Ra, ^{224}Ra, ^{226}Ra and ^{228}Ra isotopes, apart from the above-mentioned general trend at low spins, in ^{218}Ra nucleus, after steeply crossing the $\delta E(I) = 0$ line at $\sim$8 ℏ, the displacement energy becomes negative and remains so and only after spin of 25 ℏ the $\delta E(I)$ rises and then again crosses over to positive $\delta E(I)$ values. The trend for ^{220}Ra is comparatively a slower fall to the zero energy displacement line at spin $\sim$9 ℏ and then to the negative values and it again crosses this zero line at $\sim$23 ℏ to positive $\delta E(I)$ values. The whole change in the displacement values is rather smooth. The maximum negative value of displacement reached is $\delta E(I) \sim 46$ keV. In ^{222}Ra and ^{224}Ra the trend is similar. Both cross the $\delta E(I) = 0$ line at $\sim$9 ℏ. The maximum negative value of $\delta E(I)$ in ^{222}Ra is $\sim$17 keV. In ^{226}Ra, the zero crossing of displacement energy is at a larger spin value of $\sim$13 ℏ. The situation in ^{228}Ra is quite different as the $\delta E(I) = 0$ line is not crossed till the observed spin of 17 ℏ. From the above analysis of the displacement energies, it seems that the best candidates for stable octupole deformation are ^{222}Ra ($N = 134$) and ^{224}Ra ($N = 136$) nuclei.

Let us now examine the displacement energy $\delta E(I)$ versus spin I plots (Fig. 3.9, lower panel) for the even-even Th isotopes. In ^{222}Th, the energy displacement curve crosses the zero line at spin $\sim$8 ℏ, then $\delta E(I)$ becomes negative to a maximum value of $\sim$35 keV. At a spin

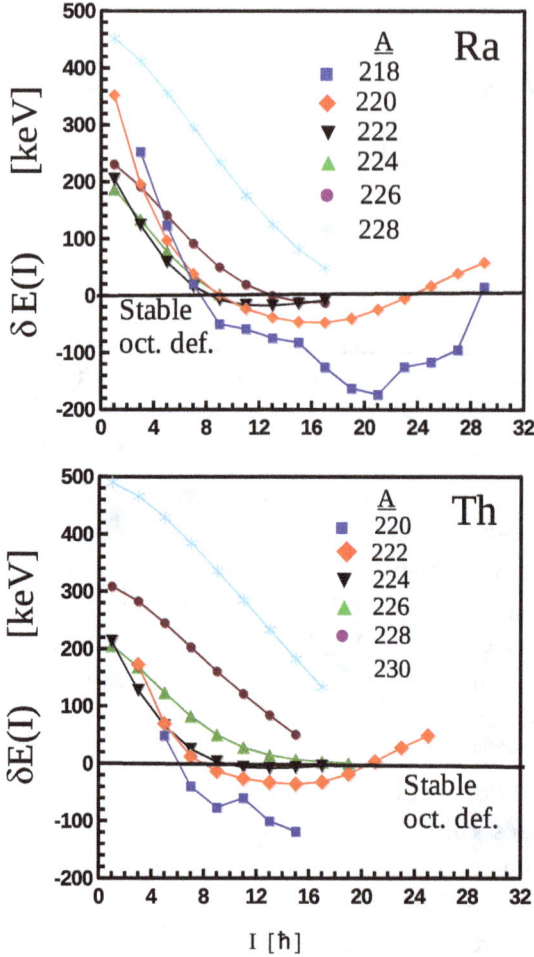

Fig. 3.9. Displacement energy or parity splitting $\delta E(I)$ between negative and positive parity states of the octupole and the ground state bands respectively, in $N = 130, 132, 134, 136, 138$ and 140 even-even Ra (upper panel) and Th (lower panel) isotopes versus spin I. The numbers in the figures refer to mass number A. The experimental data are taken from [13 and references therein].

of $\sim 21\,\hbar$, it again crosses the zero line and rises to positive $\delta E(I)$ values. In ^{224}Th, the $\delta E(I) = 0$ line is crossed at $\sim 9\hbar$ and then the energy displacement value becomes negative to a small $\delta E(I)$ of $\sim 8\,\mathrm{keV}$. In ^{226}Th, the energy displacement becomes $\delta E(I) = 0$ at

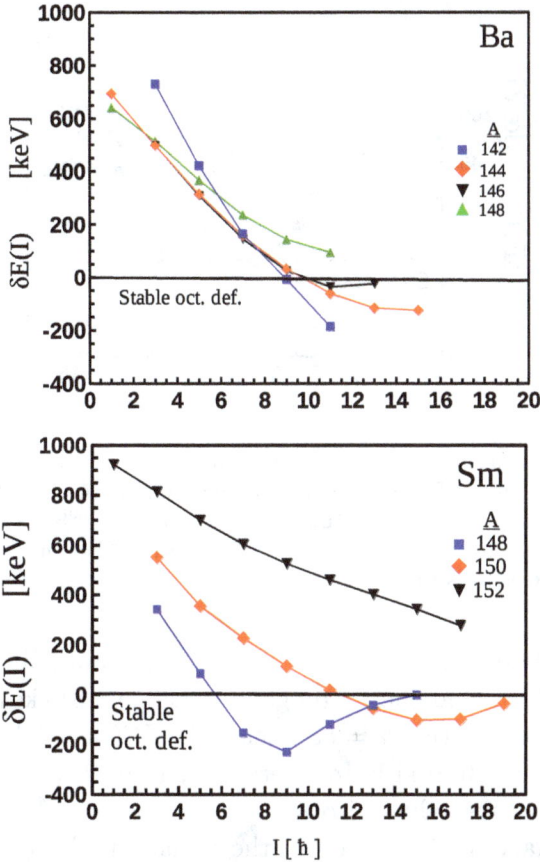

Fig. 3.10.   Displacement energy or parity splitting $\delta E(I)$ between negative and positive parity states of the octupole and the ground state bands respectively, in $N = 86$, 88, 90 and 92 even-even Ba (upper panel) and in $N = 86$, 88 and 90 even-even Sm isotopes (lower panel) versus spin I. The numbers in the figures refer to mass number A. The experimental data are taken from [13 and references therein].

spin $\sim$19 $\hbar$. In the heavier Th nuclei, ^{228}Th and ^{230}Th, the $\delta E(I) = 0$ line may be reached at still higher spins. Amongst the Th even-even isotopes, ^{224}Th appears to be a promising candidate for stable octupole deformation.

Figures 3.10 and 3.11 show plots similar to those in Fig. 3.9, but for the even-even Ba, Sm and the $N = 88$ nuclei respectively. In

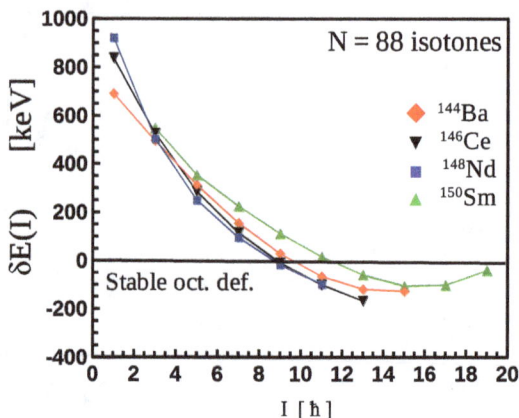

Fig. 3.11. Displacement energy or parity splitting $\delta E(I)$ between negative and positive parity states of the octupole and the ground state bands respectively, in $N = 88$ even-even isotones ^{144}Ba, ^{146}Ce, ^{148}Nd and ^{150}Sm versus spin. The experimental data for ^{146}Ce from [31] and for ^{144}Ba, ^{148}Nd and ^{150}Sm are taken from [13 and references therein].

general, in these lanthanide nuclei, the $\delta E(I)$ values at a low spin I of 3 ℏ is large in the energy range of ∼340 to ∼814 keV. In several of the Ba, Sm and the $N = 88$ nuclei, the $\delta E(I) = 0$ line is crossed between spin $I \sim 6$ to 11 ℏ. In several nuclei, the displacement energy swings to negative values also.

A summary of the trends in the variation of the displacement energy $\delta E(I)$ as a function of spin I is given in Table 3.1.

Based on the reflection asymmetric shell model, in [18], the parity splitting or the displacement energy $\delta E(I)$ for the Ra isotopes has been calculated in their work as a function of octupole deformation ε_3 and spin I, at constant values of quadrapole and hexadecapole deformation. The results of the calculations are given in their Fig. 1. It was found that displacement energy decreases with the increase in octupole deformation and with the increase in spin I. At octupole deformation $\varepsilon_3 \sim 0.07$, for ^{222,224,226}Ra, the calculated energy displacement $\delta E(I)$ at low spin of 1^- is ∼200 keV which is in agreement with the experimental data (see Fig. 3.9, upper panel).

A number of theoretical attempts [26–30], have been made to explain not only the dependence of displacement energy $\delta E(I)$ on

Table 3.1. A summary of the displacement energies $\delta E(I)$ and values of spin I at the crossings of the $\delta E(I) = 0$ line, in the even-even Ra $(Z = 88)$, Th $(Z = 90)$, Ba $(Z = 56)$, Sm $(Z = 62)$ and the $N = 88$ nuclei.

Nucleus	N	$\delta E(I)$ at $I = 3\,\hbar$ (keV)	Spin I at first $\delta E(I) = 0$ ($\hbar$)	Max. $-$ ve $\delta E(I)$ (keV)	Spin I at second $\delta E(I) = 0$ ($\hbar$)
^{220}Ra	132	196	~9	~46	~23
^{222}Ra	134	124	~9	~17	
^{224}Ra	136	134	~9		
^{226}Ra	138	192	~13	~12	
^{228}Ra	140	413	> 17		
^{222}Th	132	174	~8	~35	~21
^{224}Th	134	127	~9	~8	
^{226}Th	136	169	~19		
^{228}Th	138	283			
^{230}Th	140	467			
^{142}Ba	86	729	~9	~187	
^{144}Ba	88	497	~9	~126	
^{146}Ba	90	497	~9	~37	
^{148}Ba	92	513			
^{148}Sm	86	341	~6	~231	
^{150}Sm	88	549	~11	~105	
^{152}Sm	90	814			
^{144}Ba	88	497	~9	~126	
^{146}Ce	88	526	~9	~167	
^{148}Nd	88	504	~9	~98	
^{150}Sm	88	549	~11	~105	

angular momentum/spin I at low and high spins but also the crossings of the $\delta E(I) = 0$ line two times once at low and then a second crossing at high spin. The crossings gives rise to sign inversion of parity splitting. In [26, 27, 30], their calculations in Ra and Th isotopes were based on a one-dimensional potential well model with axial symmetry. In the work [28], a unified model based on a Hamiltonian was invoked to explain both the parity splitting at low and high angular momenta and sign inversion in parity splitting. The agreement of theory with experimental data is reasonably good. The reader is referred to these papers for detailed conclusions that were arrived at in these works.

In [32], the authors have introduced a non-dimensional characteristic (energy ratio $R_{oct}(I)$ defined in detail in their work) of the
alternating parity bands and investigated the stabilization of the
octupole deformation as a function of angular momentum in some
Ba and actinide nuclei.

Let us reiterate that in the experimental data on displacement
energy in ^{220}Ra (Fig. 3.9, upper panel) and ^{222}Th (Fig. 3.9, lower
panel), the crossings of the $\delta E(I) = 0$ line are observed at low spin
of $\sim$9 ℏ and again at high spin of $\sim$21–23 ℏ. In general, we learn from
the experimental plots of $\delta E(I)$ versus spin I (see Figs. 3.9–3.11) that
at low spins, octupole deformation is promoted with increase in spin
(or say, rotational frequency).

3.5. Octupole Vibrational and Octupole Deformed Nuclei

As mentioned earlier, octupole correlations have been found in nuclei
both in the actinide and the lanthanide regions for specific combinations of neutron and proton numbers. In this section, we will discuss
the various attempts made by researchers regarding the information
that can be derived from the observed properties of energy levels of
ground state positive parity and the octupole negative parity bands
in the even-even actinide and the lanthanide nuclei, for distinguishing
between the octupole vibrational and octupole deformed nuclei.
Specifically, the ratio of rotational frequencies in the negative and
the positive parity bands $\omega_{rot}(\pi = -)/\omega_{rot}(\pi = +)$, staggering in
the energy ratio $E(J^{\pi})/E(2^{+})$ between the positive and the negative
parity band levels, aligned angular momentum i and the difference
$\Delta i_x = [i(-) - i(+)]$ between particle aligned angular momentum
i between the negative parity and the positive parity bands, will
be considered. Also the total aligned angular momentum I_x and
kinematic moment of inertia will be discussed.

3.5.1. *Ratios of rotational frequencies* $\omega_{rot}(\pi = -)/\omega_{rot}(\pi = +)$

An indication of whether a nucleus is ocutpole deformed or octupole
vibrational, is provided by the ratio of rotational frequencies of the

negative parity and the positive parity bands. According to [23, 33]

$$\frac{\omega_{\text{rot}}(\pi = -)}{\omega_{\text{rot}}(\pi = +)} = 2\frac{E^{\pi=-}(I+1) - E^{\pi=-}(I-1)}{E^{\pi=+}(I+2) - E^{\pi=+}(I-2)} \tag{3.5}$$

This ratio is equal to unity for a perfect reflection asymmetric nucleus. In another limit, this ratio should be equal to $[4(I-3) - 2]/(4I-2)$ for the rotation of an aligned octupole phonon [24]. In Figs. 3.12 to 3.14 are plotted this ratio, as given in Eq. (3.5) above, as a function of spin I, for the even-even Ra, Th, Ba, Sm isotopes and $N=88$ isotones. In Fig. 3.12 (upper lanel), the variation is shown of the rotational frequency ratio with spin for the Ra isotopes from $N=132$ to 140. The nuclei ^{222}Ra and ^{224}Ra reach the stable octupole deformation limit of unity relatively fast, at $I \sim 12\,\hbar$. ^{226}Ra reaches this limit of unity relatively slow whereas the nucleus ^{228}Ra gives a clear indication of it being an octupole vibrational nucleus since it more or less follows the aligned octupole phonon vibrational trend. Let us now examine Fig. 3.12 (lower panel) which is a similar plot of this ratio for the even-even isotopes of Th nuclei for $N = 132$ to 142. Amongst the Th nuclei, ^{224}Th appears to be the best candidate to be an octupole deformed nucleus as the rise of the rotational frequency ratio is fast and reaches the stable octupole limit for this ratio of unity at $I \sim 12\,\hbar$, even if the nuclei ^{222}Th and ^{226}Th also attain this limit but at somewhat higher spin values. The nucleus ^{228}Th appears to demarcate the boundary between an octupole deformed and an octupole vibrational nucleus. The nuclei ^{230}Th and ^{232}Th depict octupole vibrational trend.

In Fig. 3.13 (upper panel) is plotted the ratio $\omega_{\text{rot}}(\pi = -)$ $/\omega_{\text{rot}}(\pi = +)$ as a function of spin I, for the even-even isotopes of Ba ($Z = 56$), ^{142}Ba, ^{144}Ba, ^{146}Ba and ^{148}Ba with N from 86 to 92. It is evident from this plot that in all these Ba nuclei, the ratio approaches the limit for stable octupole deformation at $I \sim 12$–$14\,\hbar$. In the lower panel of this figure are similar plots for Sm ($Z = 62$) nuclei, ^{148}Sm, ^{150}Sm and ^{152}Sm with N from 86 to 90. Here, it is clear that the nucleus ^{152}Sm ($N = 90$) follows the aligned octupole vibrational trend.

Let us now consider the even-even $N = 88$ isotones. Figure 3.14 shows a plot of the ratio $\omega_{\text{rot}}(\pi = -)/\omega_{\text{rot}}(\pi = +)$ for these nuclei

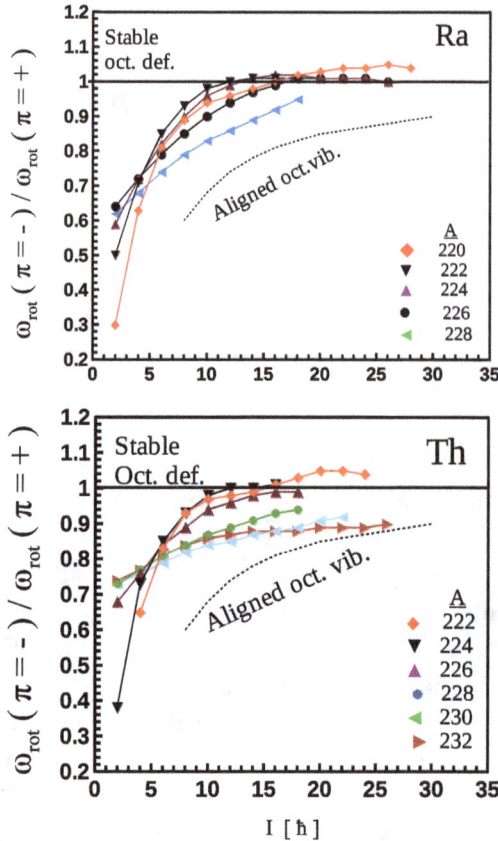

Fig. 3.12. Plots of the ratio of rotational frequencies $\omega_{\rm rot}(\pi=-)/\omega_{\rm rot}(\pi=+)$ for the negative and the positive parity bands in the even-even Ra (upper panel) from $N=132$ to 140 and Th (lower panel) isotopes from $N=132$ to 142, as a function of spin I. This ratio is unity for a reflection asymmetric rotor. The dashed curve shows the variation of this ratio with spin for aligned octupole phonon (see text). The experimental data for ^{224,226}Ra taken from [19] and ^{220,222,228}Ra and Th nuclei taken from [13 and references therein].

as a function of spin I. The isotones ^{144}Ba and ^{150}Sm do attain the value unity for this ratio at $I \sim 14-16$ ħ. The experimental data for ^{146}Ce ($Z=58$) and ^{148}Nd ($Z=60$) are not at present available for sufficiently high spins to conclude whether these nuclei are octupole vibrational or octupole deformed.

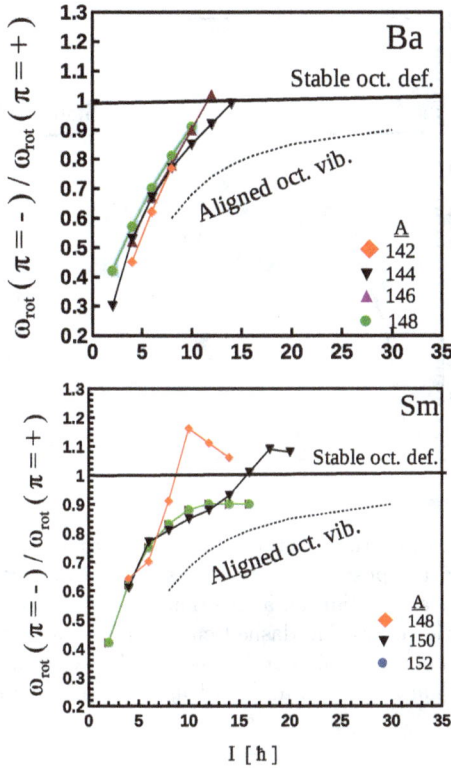

Fig. 3.13. Plots of the ratio of rotational frequencies $\omega_{rot}(\pi = -)/\omega_{rot}(\pi = +)$ for the negative and the positive parity bands in the even-even Ba (upper panel) and Sm (lower panel) isotopes, as a function of spin I. This ratio is unity for a reflection asymmetric rotor. The dashed curve shows the variation of this ratio with spin for aligned octupole phonon (see text). The experimental data taken from [13 and references therein].

A plot similar to the above plots is presented for several even-even Xe to Sm nuclei in [34], in their Fig. 6. Apart from all the other nuclei showing similar results as above, the ratio of rotational frequencies for the two Xe isotopes, ^{140}Xe and ^{142}Xe, show a trend which indicates that these Xe isotopes are octupole vibrational.

In the review article [35], and in their Fig. 20, similar plots of the ratio $\omega_{rot}(\pi = -)/\omega_{rot}(\pi = +)$ as a function of spin I, are also shown for even-even Ra, Th nuclei and for $N = 86$, 88 and 90 isotones.

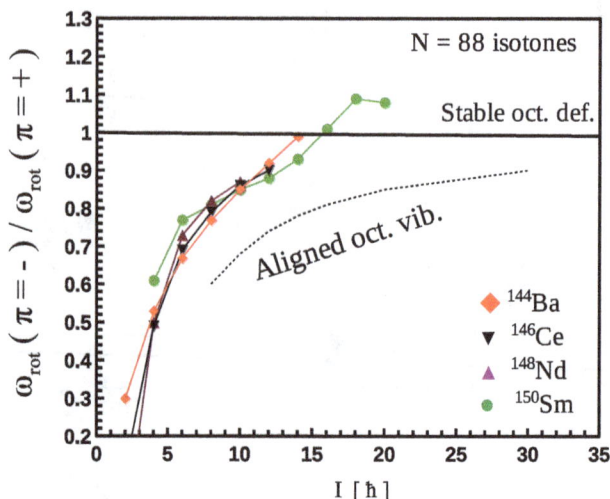

Fig. 3.14. Plot of the ratio of rotational frequencies $\omega_{rot}(\pi = -)/\omega_{rot}(\pi = +)$ for the negative and the positive parity bands in the even-even $N = 88$ isotones ^{144}Ba, ^{146}Ce, ^{148}Nd and ^{150}Sm as a function of spin. This ratio is unity for a reflection asymmetric rotor. The dashed curve shows the variation of this ratio with spin for aligned octupole phonon (see text). The experimental data for ^{146}Ce are taken from [31] and for ^{144}Ba, ^{148}Nd and ^{150}Sm are taken from [13 and references therein].

3.5.2. *Staggering in excitation energy ratios*

Let us now consider the information on nuclear shapes that can be derived from plots of experimental excitation energy ratio $E(J^\pi)/E(2^+)$ for each $J^\pi(= 1^-, 2^+, 3^-, 4^+, \ldots)$ state to that of the 2^+ first excited state as a function of spin I. For this purpose, states in the positive and the negative parity bands in the even-even Ra, Th and Ba nuclei and the $N = 88$ isotones (see Figs. 3.15–3.18) will be considered. In Fig. 3.15, these energy ratios are plotted for the even-even ^{218,220,222,224,226} and ^{228}Ra nuclei. One finds that the amplitude of the $\Delta I = 1$ energy ratio staggering between the adjacent negative and the positive parity levels, at low spins, (a) increases from lighter to heavier mass nuclei, i.e., with increase in neutron number N, (b) in the lighter mass nuclei this energy ratio staggering is almost negligible and (c) for a nucleus, it gradually decreases with increase in spin, after a certain high spin I, in general, the two rotational bands merger into a single interleaved alternating

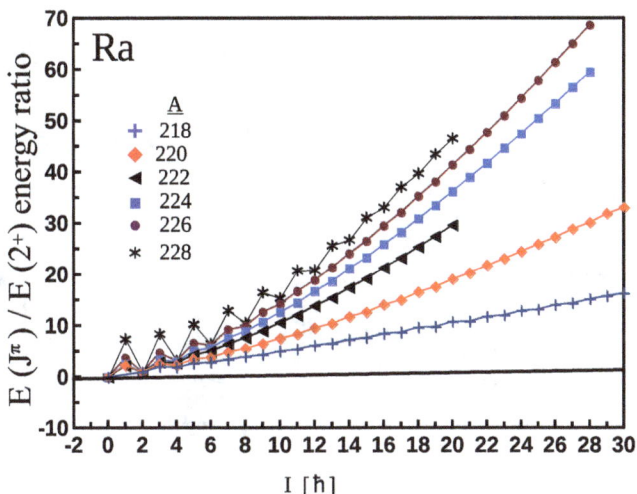

Fig. 3.15. Plots of experimental $E(J^\pi)/E(2^+)$ energy ratios versus spin I for the states in the positive parity ground state and negative parity octupole bands in even-even isotopes in ^{218,220,222,224,226,228}Ra. The experimental data for ^{224,226}Ra are taken from [19] and ^{218,220,222,228}Ra are from [13 and references therein]. These plots are very similar to those given in [6].

parity band. This phenomenon seems to be common for all the nuclei considered in this sub-section. A nucleus exhibiting octupole correlations, the observation of an interleaved alternating parity $J^\pi = 0^+, 1^-, 2^+, 3^-, 4^+, 5^-, \ldots$ band with negligible $\Delta I = 1$ energy ratio straggling, is a signature for the nucleus to attain stabilization of octupole deformed shape with increase in rotation. It is then clear from Fig. 3.15 that ^{226}Ra seems to be a borderline nucleus between the two types — octupole vibrational and octupole — deformed. Lighter nuclei than this nucleus are octupole deformed. The effect (a) mentioned above seems to be related to mixings of different quadrupole-octupole shape configurations — with increase in the quadrupole deformation parameter β_2 for the heavier Ra isotopes, the amplitude of the energy ratio staggering increases or, in other words, there are larger quantum shape fluctuations in octupole shapes at low spins for the heavier nuclei.

In Fig. 3.16 are shown plots of the experimental energy ratio $E(J^\pi)/E(2^+)$ as a function of spin I for the even-even

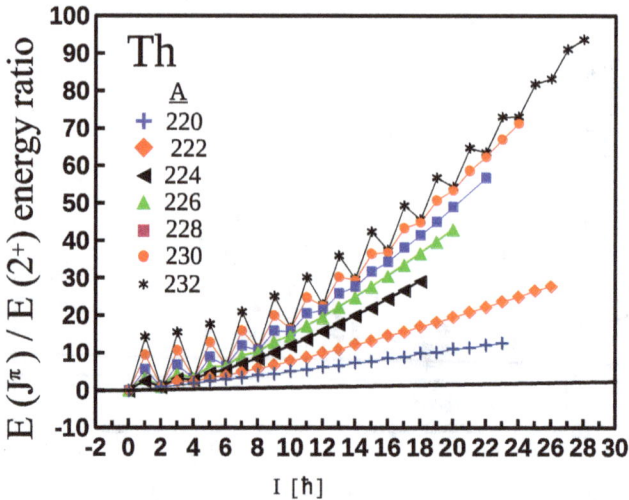

Fig. 3.16. Plots of experimental $E(J^\pi)/E(2^+)$ energy ratios versus spin I for the states in the positive parity ground state and negative parity octupole bands in even-even isotopes in ^{220,222,224,226,228,230,232}Th. The experimental data for the Th isotopes are from [13 and references therein]. These plots are very similar to those given in [6].

^{220,222,224,226,228,230,232}Th nuclei. This figure is similar to Fig. 3.15. Here also, the $\Delta I = 1$ energy ratio staggering between the adjacent negative and the positive parity levels, at low spins, exhibit phenomena as described above for the case of Ra nuclei. From a measure of this energy straggling, at low spins, one can in a similar manner, infer that ^{226}Th may be a borderline nucleus between an octupole vibrational and an octupole deformed nucleus. Nuclei lighter than this nucleus are octupole deformed.

Plots similar to those shown in Figs. 3.15 and 3.16, are displayed in Figs. 3.17 and 3.18 for the neutron-rich even-even Ba and the $N = 88$ isotones. All of these nuclei show the staggering behavior of $\Delta I = 1$ energy ratio $E(J^\pi)/E(2^+)$ as a function of spin as in the Ra and the Th nuclei. This staggering, in these lanthanide nuclei, seems to be typical of their octupole vibrational behavior.

Attempts to theoretically understand the odd-spin and even-spin excitation energy ratio $E(J^\pi)/E(2^+)$ straggling as a function of spin

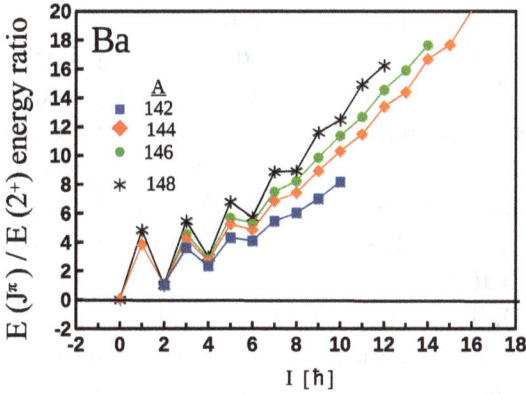

Fig. 3.17. Plots of experimental $E(J^\pi)/E(2^+)$ energy ratios versus spin I for states in the positive parity ground state and negative parity octupole bands in even-even isotopes ^{142,144,146,148}Ba. The experimental data are taken from [13 and references therein].

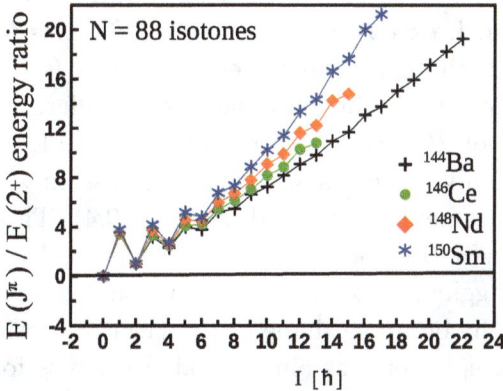

Fig. 3.18. Plot of experimental $E(J^\pi)/E(2^+)$ energy ratios versus spin for states in the positive parity ground state and negative parity octupole bands, respectively in even-even $N = 88$ isotones ^{144}Ba, ^{146}Ce, ^{148}Nd and ^{150}Sm. The experimental data for ^{146}Ce are taken from [31] and for ^{144}Ba, ^{148}Nd and ^{150}Sm are taken from [13 and references therein].

I have been done by many researchers [6, 11, 36–39, and references therein], in the actinide and the lanthanide nuclei.

In [36], symmetry-conserved beyond mean-field calculations of the above-mentioned energy ratio straggling in ^{224}Ra nucleus were carried out with the multireference relativistic energy density functional method. This study revealed some very interesting results. Calculations of octupole ($\lambda = 3$) deformation parameter β_3 for the odd-spin negative parity and the even-spin positive parity states up to spin $J = 10\,\hbar$ were done in this nucleus. It was found that for the negative parity states the octupole deformation remained almost constant (around 0.144) as a function of spin, whereas for the positive parity states, it increases from 0.101 at $J = 0$ to 0.141 at $J = 10$ (see their Fig. 3(c)). This study has shown that the spin dependent odd-even energy ratio staggering at low spins is related to the rotation induced octupole shape stabilization of the positive parity states due to the gradual drift of their octupole shape with increasing spin towards the octupole shape of the negative parity states. Calculations on similar lines as in ^{224}Ra [36], were carried out in [37], for the even-even neutron-rich Ba isotopes (^{142,144,146,148,150}Ba). They found that (1) the calculated straggling in the energy ratio $E(J^\pi)/E(2^+)$ as a function of spin I, except in the case of ^{142}Ba, was found to be in agreement with the experimental energy ratios (see their Fig. 1), (2) from detailed studies in ^{144}Ba nucleus for different combinations of the values of β_2 and β_3, e.g., for $\beta_3 = 0.1$, the straggling amplitude roughly increases with the increase in the quadrupole deformation parameter $\beta_2(= 0.0, 0.1, 0.2, 0.3, 0.4)$. This is also found experimentally in (a) above, in the case of the Ra isotopes, i.e., increase in straggling amplitude in nuclei from lighter to the heavier ones, and (3) the octupole shape of the negative parity states with the increase in spin remains almost stable whereas for the positive parity states it drifts from weakly octupole to that of the negative parity states at higher spins, as in the case of ^{224}Ra. The rotation induced octupole shape stabilization in the positive parity states can be inferred as a common phenomenon in all the actinide and the lanthanide nuclei under discussion.

In [6], a theoretical model called an analytic quadrupole octupole axially symmetric model (AQOA), was developed. It was used to calculate the above-mentioned straggling in the energy ratios in the even-even Ra and Th nuclei. The calculations use a parameter ϕ_0, which defines the relative contribution of quadrupole and octupole deformations. The theoretical energy ratio $E(J^\pi)/E(2^+)$ calculated

as a function of spin I, for ^{226}Ra using the parameter $\phi_0 = 56^0$
and for ^{226}Th $\phi_0 = 60^0$ respectively, are in excellent agreement with
the experimental data. (These results are shown in their Fig. 2).
On the basis of negligible straggling of the $(J^\pi)/E(2^+)$ ratios at
low spins, it is inferred in the work that both ^{226}Ra and ^{226}Th
nuclei are very close to the transition region between the octupole
deformed and octupole vibrational regimes. Here, it is interesting
to note that in theoretical work [40] based on a different model
which takes into account both the octupole deformation and stable
quadrupole deformation for nuclei close to axial symmetry but with
reflection asymmetry, a conclusion has been arrived at similar to
that in [6], that the nucleus ^{226}Th sits on the border line between
octupole vibrational and octupole deformed shapes. A systematic
investigation of the quadrupole and octupole collective states in
the isotopic chains in Ra, Th, Sm and Ba nuclei was done in
the work [11] using the microscopic density functional theory. The
calculated values of energy ratio $E(J^\pi)/E(2^+)$ as a function of spin I
were then compared with experimental data. (The results are shown
in their Fig. 13). It is found that straggling of $\Delta I = 1$ energy ratio
$E(J^\pi)/E(2^+)$ between the adjacent negative and the positive parity
levels is negligible for the Ra and Th isotopes which are lighter than
^{226}Ra and ^{226}Th, respectively.

As for the Ra and Th isotopes, in [11], theoretical calculations
have also been done for the energy ratio $E(J^\pi)/E(2^+)$ versus spin I
for the even-even Ba nuclei and compared with experimental data.
The calculations show $\Delta I = 1$ odd-even staggering trends similar to
those observed and are also in fair agreement with experimental data.

3.5.3. *Aligned angular momentum i and difference Δi_x*

Let us now consider the experimentally deduced quantities, the
aligned angular momentum i (or i_x) and the difference between
aligned angular momentum Δi_x, between the negative parity and
the positive parity bands, $\Delta i_x = [i(-) - i(+)]$, as a function of
rotational frequency $\hbar\omega$. The quantities $i(-)$ and $i(+)$ are the aligned
angular momenta for the negative and the positive parity bands, at
the same rotational frequency. The aligned angular momentum i or

i_x is obtained from the experimental data on a level scheme using the prescription given in [41–43]. This is done to facilitate the comparison of experimental data with theoretical predictions.

Here, we will consider $K = 0$ ground state positive parity and the negative parity bands in some of the even-even actinide and the lanthanide nuclei which exhibit octupole correlations. These bands form alternating parity bands, as discussed earlier in Sec. 3.3.

We first consider the aligned angular momentum i (or i_x). In Fig. 3.19 [19], are presented plots of aligned angular momentum i_x, as a function of rotational frequency $\hbar\omega$, for the even-even ^{220,222,224,226,228}Ra ($N = 132$ to 140) (left panel) and

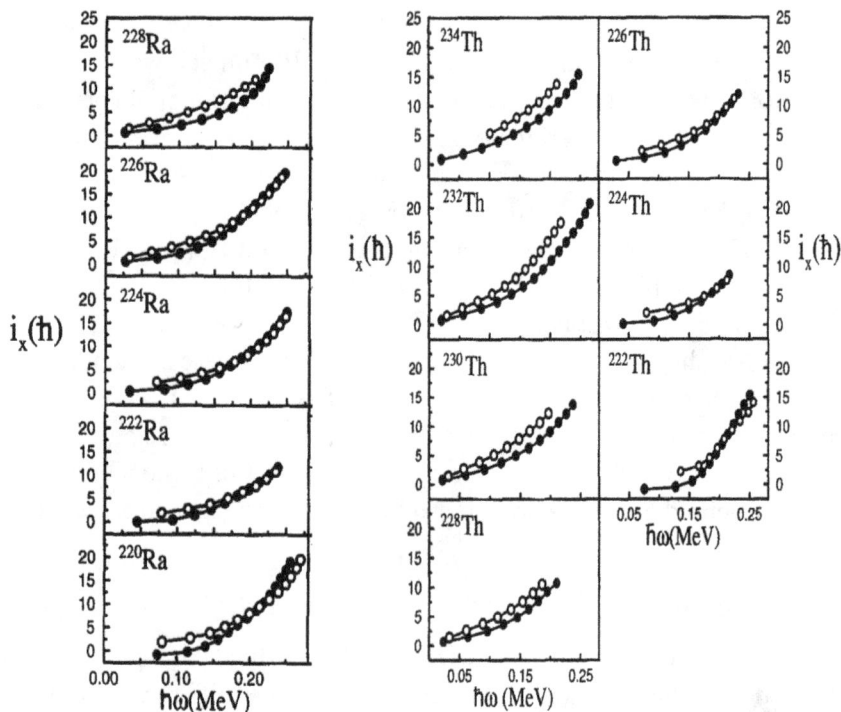

Fig. 3.19. Plots of aligned angular momentum i_x as a function of rotational frequency $\hbar\omega$ for the even-even Ra isotopes ($N = 132$ to 140) in the left panel and for the Th isotopes ($N = 132$ to 144) in the right panel. The figures are reproduced with permission from [19].

^{222,224,226,228,230,232,234}Th isotopes ($N = 132$ to 144) (right panels). It is observed that the aligned angular momentum i_x increases smoothly as a function of rotational frequency $\hbar\omega$ in all these nuclei, for both the negative and the positive parity bands. It is, therefore, apparent that, up to the high spins or rotational frequencies at which the nuclei were investigated, there are no observable rotational alignment effects.

Figure 3.20 shows plots of aligned angular momentum, i versus rotational frequency, $\hbar\omega$ for the positive parity ground state and the negative parity bands in ^{144}Ba and ^{146}Ba as observed in [21]. These two nuclei which are only two neutrons apart show different alignment patterns. In ^{144}Ba, the ground state band tends to gradually bend upwards after a rotational frequency of $\sim$0.3 MeV and in the negative parity band, there is no up bending. In ^{146}Ba, in the ground state positive parity band, at rotational frequency of ≈ 0.28 MeV, a pronounced bandcrossing with a large alignment gain of $> 8\hbar$, indicative of neutron $i_{13/2}$ pair alignment, is observed. A similar observation was made in [44]. The alignment plot of the negative parity band in this nucleus is similar to that seen in ^{144}Ba but at rotational frequency of ≈ 0.29 MeV, it is crossed by a negative parity two- quasiparticle side band based on the 1944.9 keV 7^- level.

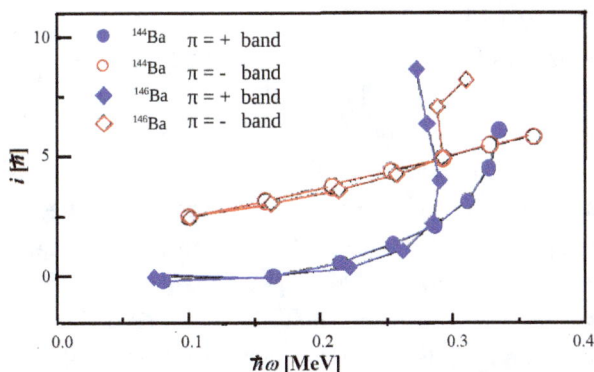

Fig. 3.20. Plots of aligned angular momentum, i versus rotational frequency, $\hbar\omega$ for the positive and the negative parity bands in ^{144}Ba and ^{146}Ba. In the calculations, the Harris parameters $J_0 = 19$ MeV$^{-1}\hbar^2$ and $J_1 = 80$ MeV$^{-3}\hbar^4$ were used. Figure is adopted from [21].

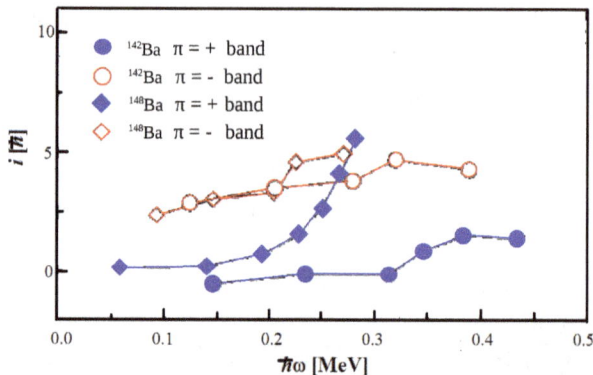

Fig. 3.21. Plots of aligned angular momentum, i versus rotational frequency, $\hbar\omega$ for the positive and the negative parity bands in ^{142}Ba and ^{148}Ba. In the calculations, the Harris parameters $J_0 = 12\,\text{MeV}^{-1}\,\hbar^2$, $J_1 = 60\,\text{MeV}^{-3}\hbar^4$ for ^{142}Ba and $J_0 = 22\,\text{MeV}^{-1}\,\hbar^2$, $J_1 = 80\,\text{MeV}^{-3}\,\hbar^4$ for ^{148}Ba, were used. Figure is adopted from [21].

Figure 3.21 similarly show plots of aligned angular momentum, i versus rotational frequency, $\hbar\omega$ for the positive parity ground state and the negative parity bands in ^{142}Ba and ^{148}Ba as observed in [21]. The alignment behavior in the positive parity ground state band in ^{148}Ba is similar to that in ^{144}Ba except that it bends upwards at a lower rotational frequency of $\sim$0.23 MeV.

Now, we try to look into the information that can be obtained from the behavior of the difference quantity Δi_x, between the positive and the negative parity bands, as a function of rotational frequency $\hbar\omega$. We follow the arguments put forward in [19, 45, 46]. One situation is when the component of the angular momentum aligned to the rotational axis for the negative parity state $i(-)$ or that for the positive parity state $i(+)$, at the same rotational frequency, is equal to the rotational angular momentum R. Then, in this case $\Delta i_x = [i(-) - i(+)] = 0$. This will occur for a permanent octupole deformed nucleus. The other situation could be when the negative parity state is formed by octupole vibrations of the rotating quadrupole deformed nucleus by the coupling of rotational angular momentum R and the angular momentum of the octupole phonon ($3\,\hbar$). If the phonon angular momentum ($3\,\hbar$) is aligned with the rotational angular momentum R, then the quantity $\Delta i_x = 3\hbar$. This will be the case when the nucleus is octupole vibrational.

In Fig. 3.22, are plots of $\Delta i_x = [i(-) - i(+)]$, as a function of rotational frequency $\hbar\omega$, for the even-even ^{218,220,222}Rn ($N = 132$, 134 and 136) (top panel), ^{220,222,224,226,228}Ra ($N = 132$ to 140) (middle panel) and ^{222,224,226,228,230}Th isotopes ($N = 132$ to 140) (bottom panel). The values of Δi_x are calculated by subtracting from the value of aligned angular momentum $i(-)$ for the negative parity state, an interpolated value $i(+)$ for the positive parity band at the same rotational frequency $\hbar\omega$ as that for the negative parity band.

The ^{222}Ra $N = 134$ and ^{224}Ra $N = 136$ isotopes behave in a similar manner throughout the rotational frequency range. The highest value of Δi_x attained is just less than $2\,\hbar$ at $\hbar\omega \sim 0.10\,\text{MeV}$. It then falls in a regular manner with the increase in rotational frequency. At $\hbar\omega \sim 0.20\,\text{MeV}$, Δi_x becomes nearly zero and then even falls below the zero line. The variation of Δi_x throughout the rotational frequency range for ^{226}Ra $N = 138$ is also similar to the ^{222}Ra $N = 134$ and ^{224}Ra $N = 136$ isotopes, except that Δi_x in this case rises to somewhat above $2\hbar$ at $\hbar\omega \sim 0.12\,\text{MeV}$. In the three isotopes ^{222,224,226}Ra with $N = 134, 136$ and 138, octupole deformation stabilizes at high rotational frequency. The $N = 132$ ^{220}Ra isotope at low rotational frequency has $\Delta i_x \sim 3\,\hbar$ and therefore, it can be considered to be an octupole vibrator at low spins. The Δi_x value, then decreases with the increase in rotational frequency up to $\hbar\omega \sim 0.21\,\text{MeV}$ when it becomes zero and then further decreases to negative values. This behavior at low rotational frequency is similar to the ^{218,220,222}Rn isotopes (see top panel). However, all three Rn isotopes considered behave like octupole vibrators ($\Delta i_x \sim 3\,\hbar$) in almost the entire rotational frequency range. The nucleus ^{228}Ra ($N = 140$) behaves differently, the Δi_x value is small (~ 0) at low rotational frequency and then it increases with the increase in frequency and reaches a value of $3\,\hbar$ at $\hbar\omega \sim 0.17\,\text{MeV}$ at which the octupole phonon gets fully aligned with the rotational axis and the nucleus behaves as an octupole vibrator.

The bottom panel in Fig. 3.22 shows plots of Δi_x as a function of rotational frequency $\hbar\omega$ for the even-even ^{222,224,226,228,230}Th isotopes

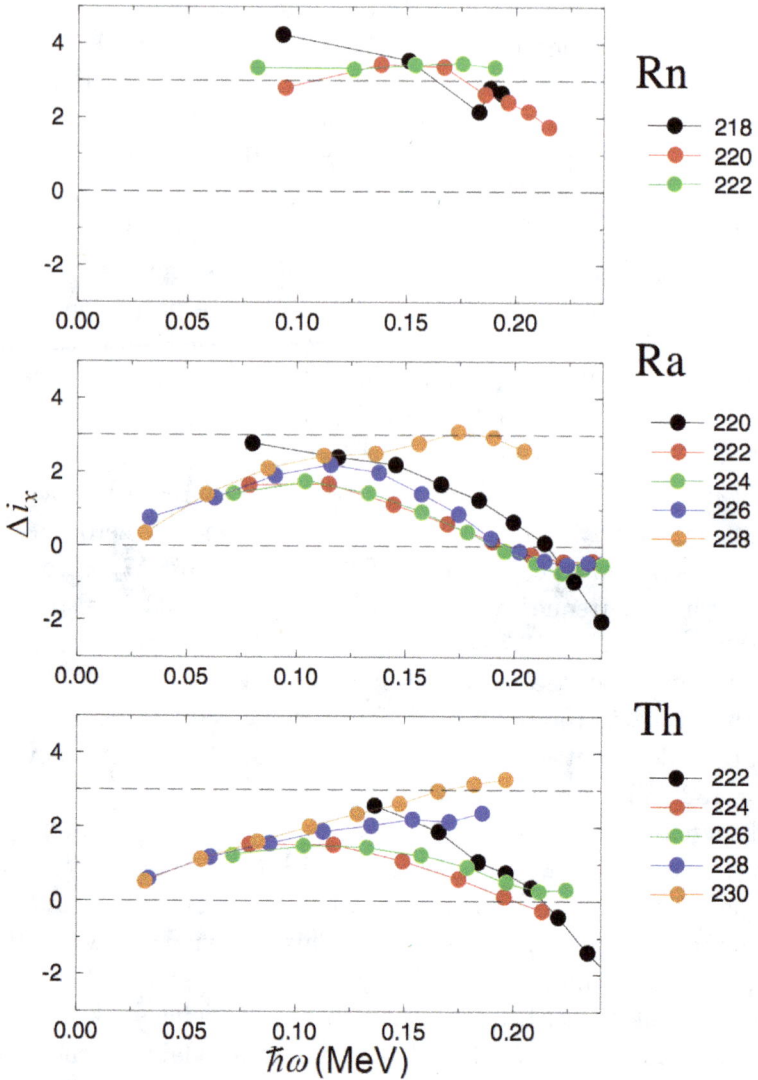

Fig. 3.22. The plots show the difference in aligned angular momentum $\Delta i_x = i(-) - i(+)$ in [$\hbar$] between the negative parity and the positive parity bands at the same rotational frequency in the even-even Rn, Ra and Th nuclei, as a function of rotational frequency, $\hbar\omega$. The dashed lines at 0 $\hbar$ and 3 $\hbar$ are the octupole rotational and the octupole vibrational limits of Δi_x respectively (see text for explanation). The figures is courtesy of Professor P.A. Butler.

(N = 132 to 140). The isotopes ^{224}Th and ^{226}Th with N = 134 and 136 respectively, behave in a manner similar to the N = 134 and 136 counterparts ^{222}Ra and ^{224}Ra. The Δi_x values are smaller than 2 ℏ and decrease after ℏω ~ 0.10 MeV with the increase in rotational frequency to ~0 at ℏω ~ 0.21 MeV where stable octupole deformation is fully developed. The N = 140 isotones ^{230}Th and ^{228}Ra behave similarly. The Δi_x value is small at low frequencies and it increases gradually with rotational frequency till the octupole phonon is fully aligned, with alignment gain (Δi_x) being ~3 ℏ at rotational frequency ℏω ~ 0.17 MeV. The ^{228}Th nucleus with N = 138 behaves as a transitional nucleus between the octupole deformed ^{226}Th and octupole vibrational ^{230}Th. The N = 132 isotones ^{222}Th and ^{220}Ra behave somewhat similarly as the rotational frequency increases. It is pertinent to mention here that there are structural shape differences between these two nuclei at very high spin 25 ℏ onwards [16].

Figure 3.23, shows plots similar to those in Fig. 3.22, for the even-even Xe, Ba, Ce, Nd and the Sm nuclei. From these plots, it is apparent that most nuclei in this region conform to qualify as octupole vibrators in a large part of the rotational frequency region. However, at high rotational frequencies, N = 88 ^{144}Ba, and the N = 90 isotones, ^{148}Ce and ^{150}Nd exhibit trends of being octupole deformed.

3.6. Total Aligned Angular Momentum and Kinematic Moment of Inertia

In this section, the rotational behavior of axially symmetric and reflection asymmetric pear-shaped vibrational and deformed nuclei in the region of light even-even actinides and the lanthanides, is discussed. For this purpose, plots of total aligned angular momentum I_x and kinematic moment of inertia, $\mathcal{I}^{(1)} (= I_x/\omega)$ as a function of rotational frequency ℏω, are considered. The rotational frequency and the total aligned angular momentum were calculated following the usual prescriptions as described in [41–43, 47, 48]. The behavior of aligned angular momentum i_x and the difference in aligned

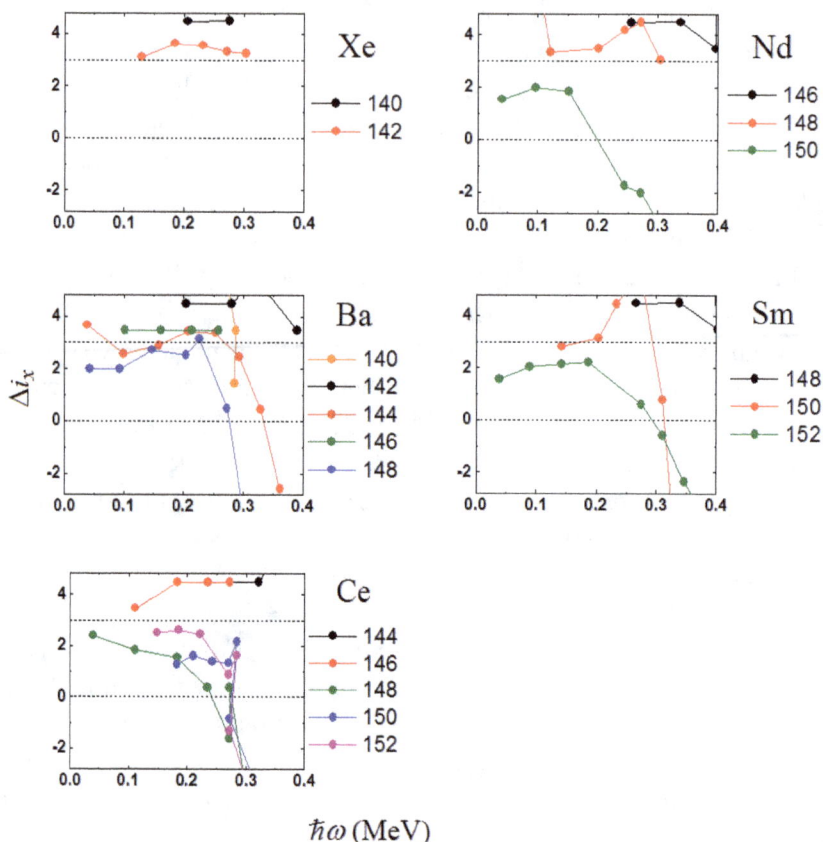

Fig. 3.23. Plots similar to those shown in Fig. 3.22, but for the even-even Xe, Ba, Ce, Nd and Sm even-even nuclei, as a function of rotational frequency, $\hbar\omega$. The dotted lines at $0\,\hbar$ and $3\,\hbar$ are the octupole rotational and the octupole vibrational limits of Δi_x respectively (see text for explanation). The figure is courtesy of Professor P.A. Butler.

angular momentum Δi_x have already been discussed in the preceding Section 3.5.3.

In Figure 3.24, plots of total aligned angular momentum I_x as a function of rotational frequency $\hbar\omega$ are shown, for the $N = 130$ isotones ^{218}Ra and ^{220}Th. An irregular zig-zag behavior around nearly a constant rotational frequency is exhibited which indicates that these nuclei seem to be soft quadrupole vibrators or nuclei with

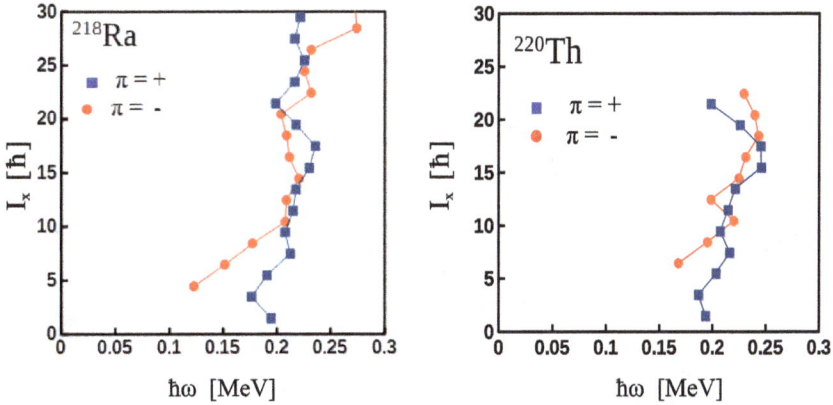

Fig. 3.24. Plots of total aligned angular momentum, I_x versus rotational frequency, $\hbar\omega$, for the yrast +ve parity ground state bands and the −ve parity octupole bands in the $N = 130$ isotones ^{218}Ra and ^{220}Th. The experimental data are taken from [13 and references therein].

low collectivity. The $E(4^+)/E(2^+)$ ratios for these nuclei are 1.90 and 1.97 respectively.

Figure 3.25 shows the level schemes of the $N = 132$ isotones ^{220}Ra and ^{222}Th [16]. In these two level schemes, a contrasting behavior is seen. In ^{220}Ra, the positive parity band is seen up to $J^\pi = 28^+(30^+)$ and the negative parity band up to $J^\pi = 29^-(31^-)$ whereas in ^{222}Th these bands are observed only up to $J^\pi = 24^+$ and $J^\pi = 23^-$ (25^-). Figure 3.26 shows plots of total aligned angular momentum, I_x as a function of rotational frequency, $\hbar\omega$, for the yrast positive parity ground state bands and the negative parity octupole bands in these two isotones. The observation of a rather regular increase in I_x with rotational frequency $\hbar\omega$, in both these nuclei, indicate an increase in quadrupole collectivity in comparison to the $N = 130$ isotones. The $E(4^+)/E(2^+)$ ratios for these nuclei are 2.29 and 2.40, respectively. For ^{222}Th nucleus $\beta_2 = 0.153$ [17].

Figure 3.27 plots of kinematic moment of inertia, $\mathcal{J}^{(1)}$, as a function of rotational frequency, $\hbar\omega$, are shown, for the yrast positive parity ground state bands and the negative parity octupole bands in the $N = 132$ isotones ^{220}Ra and ^{222}Th. In both the nuclei, ^{220}Ra and ^{222}Th, a bandcrossing is indicated in the negative parity bands

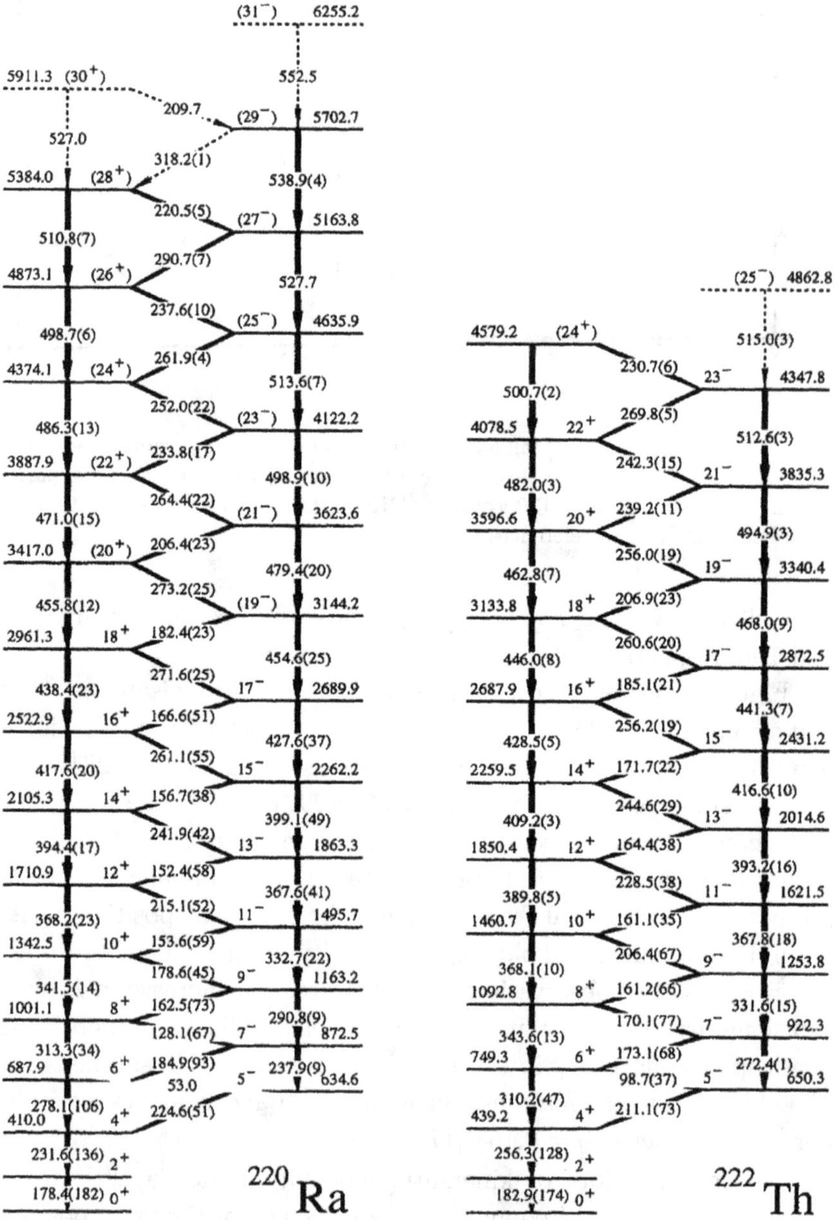

Fig. 3.25. Level schemes of ^{220}Ra and ^{222}Th [16]. Figure is reproduced with permission from [16].

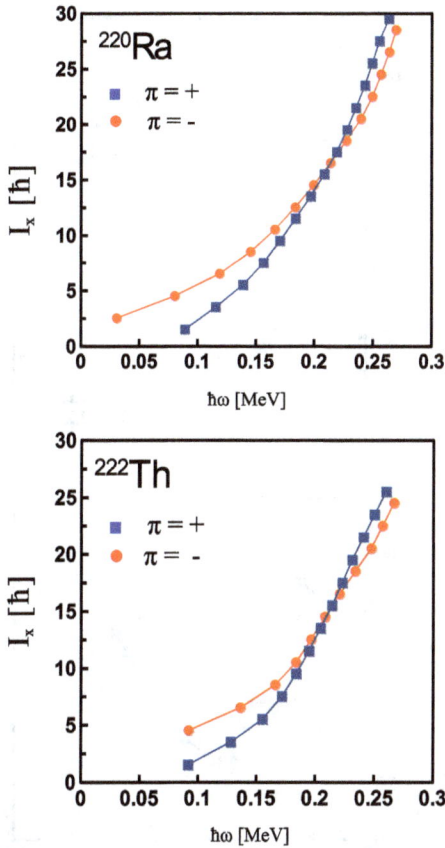

Fig. 3.26. Plots of total aligned angular momentum, I_x, as a function of rotational frequency, $\hbar\omega$, for the yrast positive parity ground state bands and the negative parity octupole bands in the $N = 132$ isotones ^{220}Ra and ^{222}Th. The experimental data are taken from [16] and [13 and references therein].

at $\hbar\omega \sim 0.21$ MeV which may be due to the $\nu(j_{15/2})^2$ neutron pair aligned band as predicted by theory [33] (see the paragraph below).

Detailed microscopic calculations were carried out in the cranked reflection-asymmetric Woods–Saxon–Bogolyubov-Strutinsky framework in some of the even-even Ra and Th nuclei in [33] to study the evolution of the nuclear shape with rotational frequency/angular momentum. It was found that in nuclei around ^{220}Ra, the reflection-asymmetric shape should persist till very high

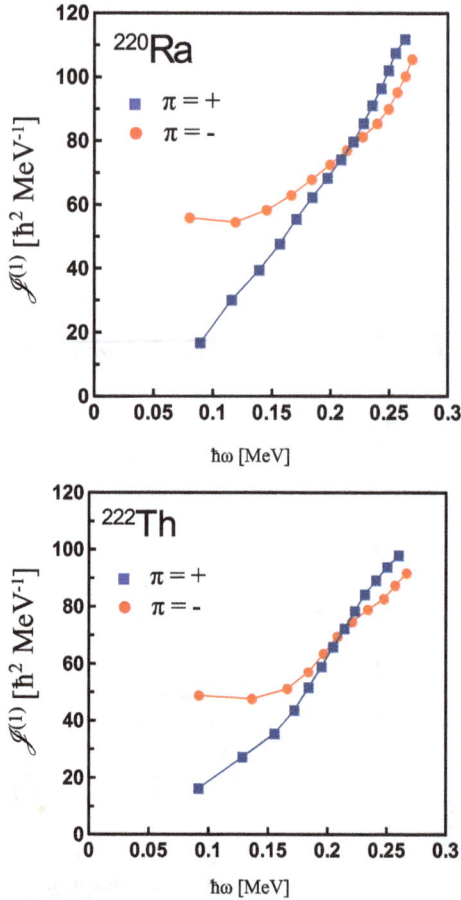

Fig. 3.27. Plots of kinematic moment of inertia, $\mathcal{J}^{(1)}$, as a function of rotational frequency, $\hbar\omega$, for the yrast positive parity ground state bands and the negative parity octupole bands in the $N = 132$ isotones ^{220}Ra and ^{222}Th. The experimental data are taken from [16] and [13 and references therein].

spins. In ^{222}Th, the calculations were extended to include particle-number projection with pairing. The calculations performed for the reflection-asymmetric shape ($\beta_2 = 0.116$ and $\beta_3 = 0.104$) showed that the octupole deformed ground-state band crosses a $\nu(j_{15/2})^2$ neutron pair aligned band at rotational frequency of $\hbar\omega \sim 0.20$ MeV with large interaction. For ($\beta_2 = 0.120$ and $\beta_3 = 0$) reflection-symmetric shape, the spin alignments are large and the interaction between the crossing bands small. The quasiparticle aligned neutron pair configuration $\nu(j_{15/2})^2$ and the proton pair $\pi(i_{13/2})^2$ configuration become the lowest already at spin $\sim 12\,\hbar$. The reflection-asymmetric configuration remains yrast until spin $\sim 24\,\hbar$ and then the four-quasiparticle configuration $\nu(j_{15/2})^2\pi(i_{13/2})^2$ becomes yrast at spin $\sim 26\,\hbar$ when a transition in the nuclear shape is predicted from reflection-asymmetric to reflection-symmetric shape. The results of these calculations in ^{222}Th are shown in Fig. 12 in [33].

The absence of the bands in ^{222}Th in [16] above spin $\sim 24\,\hbar$ can be attributed to the crossing of aligned quasiparticle bands which causes the gamma intensity flow to be divided between a couple of states near the band-crossing region thus the de-exciting connecting gamma-ray transitions become too weak to be observed experimentally above spins 24^+ and 25^-.

In Fig. 3.28 are shown plots of kinematic moment of inertia $\mathscr{J}^{(1)}(=I_x/\omega)$ as a function of rotational frequency $\hbar\omega$, for the even-even $N = 134$ to 138 ^{222}Ra, ^{224}Ra, ^{226}Ra (left panels) and ^{224}Th, ^{226}Th, ^{228}Th (right panels) nuclei. In all these nuclei, for the positive parity ground state bands, the kinematic moment of inertia increases with the increase in rotational frequency. In the negative parity bands, there is an initial near constancy of the moment of inertia at low spins and then at higher spins the moment of inertia increases with the increase in rotational frequency. Another common behavior seen in these plots is that the kinematic moment of inertia at low rotational frequencies, in the negative parity bands, is higher than for the positive parity bands. The kinematic moment of inertia difference between the opposite parity bands decreases with the increase of rotational frequency. This situation is valid for ^{220}Ra,^{222}Ra, ^{224}Ra, ^{226}Ra and ^{222}Th,^{224}Th, ^{226}Th nuclei (see Fig. 3.27 for ^{220}Ra and ^{222}Th nuclei). The moment of inertia merges to a common value at a certain value of rotational frequency ($\sim 0.2\,$MeV) and then it nearly remains so for higher rotational frequencies, except, as mentioned earlier, for ^{220}Ra and ^{222}Th nuclei where quasi-particle alignments occur at high rotational frequencies. This behavior is an indication of the stabilization of the octupole

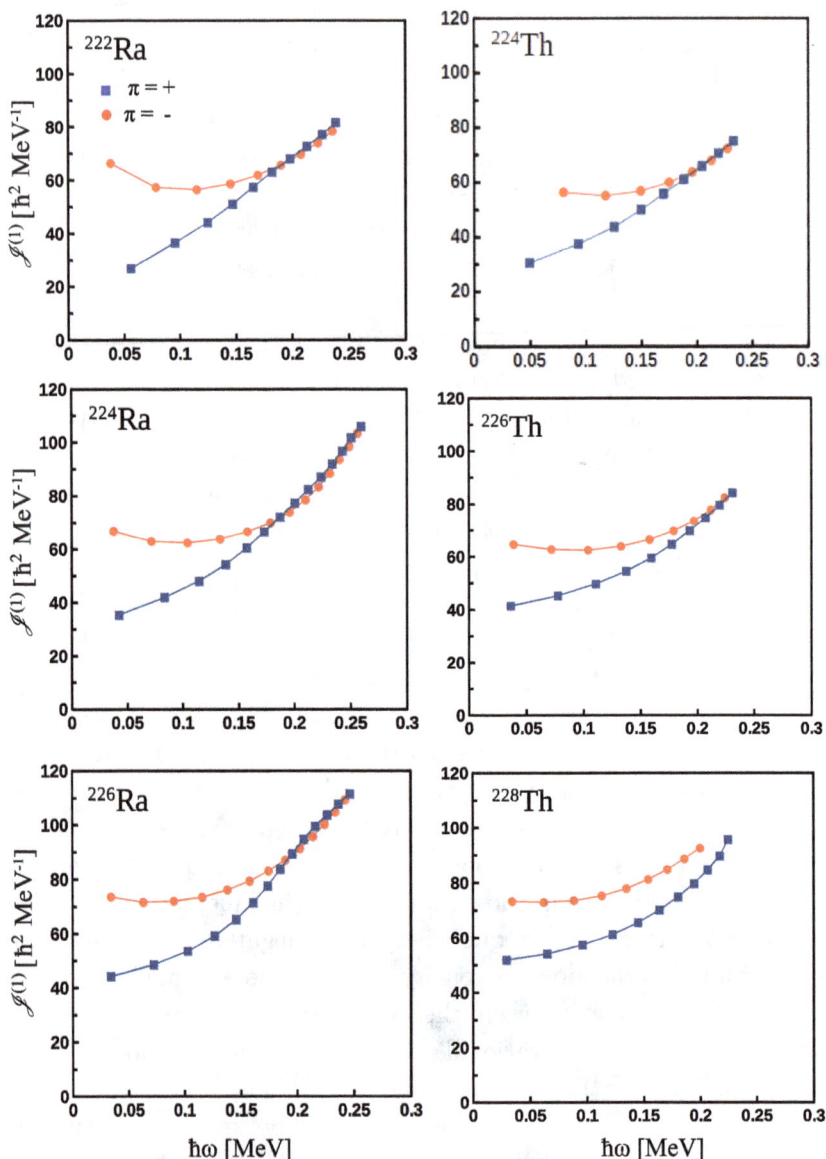

Fig. 3.28. Plots of kinematic moment of inertia, $\mathscr{I}^{(1)}$ as a function of rotational frequency, $\hbar\omega$ for the positive and the negative parity bands in the even-even $N = 134$ to 138 ^{222}Ra, ^{224}Ra, ^{226}Ra (left panels) and ^{224}Th, ^{226}Th, ^{228}Th (right panels) isotopes. The experimental data for ^{224,226}Ra are taken from [19] and for the other nuclei from [13 and references therein].

deformed shape as the rotational frequency increases. However, for ^{228}Th ($N = 138$), although there is a decrease in the difference of moment of inertia between the negative and the positive parity bands with the increase of rotational frequency, the two bands do not attain a common value until the presently observed high spin states.

The self-consistent Hartree–Fock calculations in [49] with quadrupole-octupole degrees of freedom suggest a possible explanation for the kinematic moment of inertia being higher at low rotational frequencies in the negative parity bands as compared to these in the positive parity bands, in terms of larger quadrupole deformation for the negative parity states. According to [50], this feature is not predicted in the Strutinski-type theoretical calculations. Another explanation for higher kinematic moment of inertia in the negative parity band because of higher total aligned angular momentum I_x for the negative parity band, has been provided [51] in terms of one octupole phonon alignment in this band (e.g., in ^{220}Ra and ^{222}Th at $\hbar\omega \sim 0.1\,\mathrm{MeV}$, $I_x(-) - I_x(+)$ is $\sim 3\,\hbar$; see Fig. 3.26). The decrease in the difference of kinematic moments of inertia between the negative and the positive parity bands with increasing rotational frequency as a result of the decrease in the quantity $\{I_x(-) - I_x(+)\}$, has been explained by successive mixing of alignment of octupole phonons — see [51] for details.

Let us now consider the behavior of kinematic moment of inertia $\mathcal{J}^{(1)}$ as a function of rotational frequency $\hbar\omega$, in the even-even ^{142}Ba$_{86}$, ^{144}Ba$_{88}$, ^{146}Ba$_{90}$ and ^{148}Ba$_{92}$ isotopes (see Fig. 3.29) and in the $N = 88$ isotones ^{144}Ba, ^{146}Ce, ^{148}Nd and ^{150}Sm (see Fig. 3.30). In all these nuclei, a common feature is the enhanced kinematic moment of inertia at low rotational frequencies in the negative parity bands in comparison to that in the positive parity bands. This property is similar to that found for the even-even actinides Ra and Th (see Figs. 3.27 and 3.28). In ^{146}Ba (see also Fig. 3.20), the ground state positive parity band exhibits an upbending at $\hbar\omega \sim 0.29\,\mathrm{MeV}$ which is characteristic of band crossing of a two quasiparticle band. This is

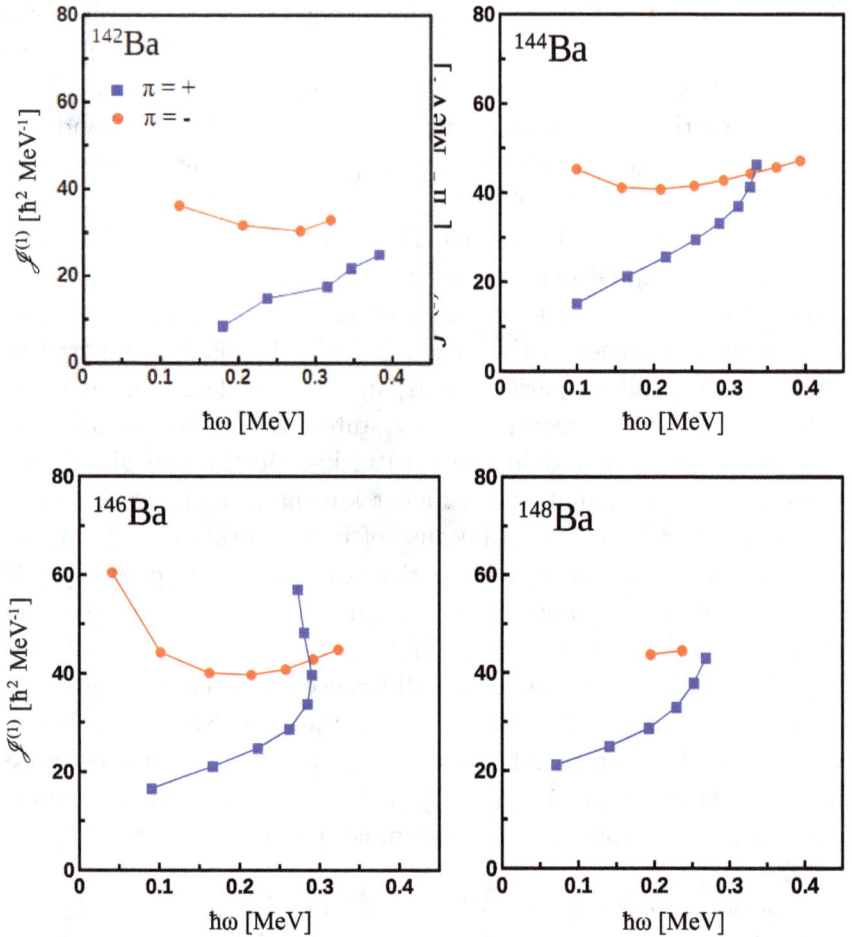

Fig. 3.29. Plots of kinematic moment of inertia, $\mathscr{J}^{(1)}$ as a function of rotational frequency, $\hbar\omega$ for the positive and the negative parity bands in the even-even ^{142}Ba$_{86}$, ^{144}Ba$_{88}$, ^{146}Ba$_{90}$ and ^{148}Ba$_{92}$ isotopes. The experimental data are taken from [13 and references therein]. In addition, see [52] for additional data on ^{146}Ba nucleus.

also an indication of shape transition from reflection-asymmetric to reflection-symmetric in this nucleus. In ^{144}Ba (also see Fig. 3.20), the experimental data is not available to know if a delayed bandcrossing occurs at $\hbar\omega \sim 0.34$ MeV.

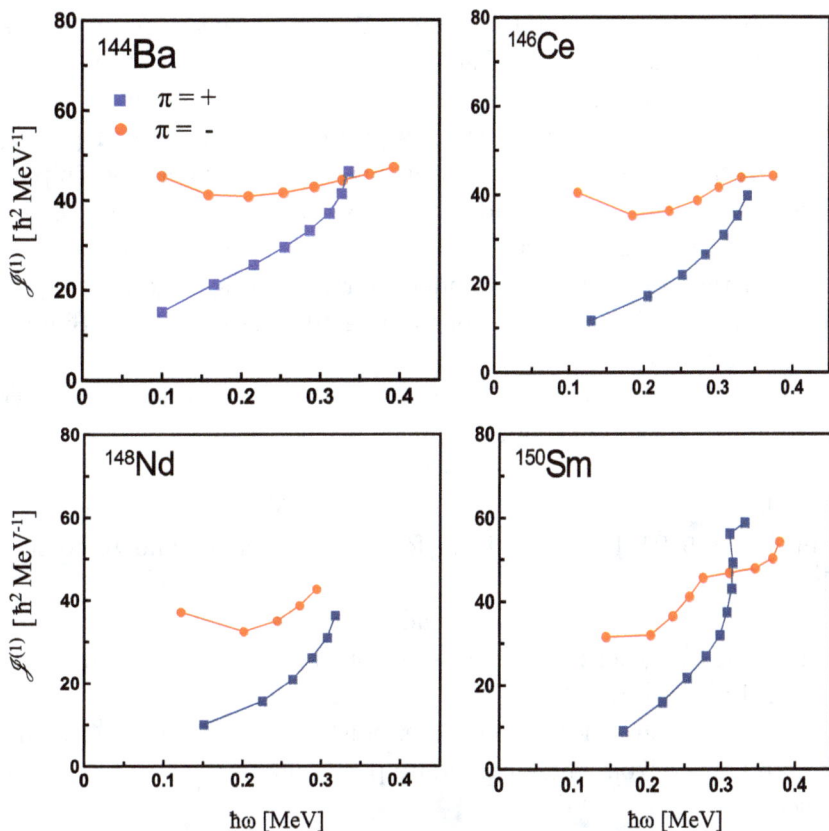

Fig. 3.30. Plots of kinematic moment of inertia, $\mathcal{J}^{(1)}$ as a function of rotational frequency, $\hbar\omega$ for the positive and the negative parity bands in the even-even $N = 88$, ^{144}Ba, ^{146}Ce, ^{148}Nd and ^{150}Sm isotones. The experimental data for ^{146}Ce are taken from [31], for ^{150}Sm from [53, 54] and for ^{144}Ba and ^{148}Nd taken from [13 and references therein].

B. Odd Mass Nuclei

3.7. Yrast Parity Doublets

In the theoretical calculations of [55], in the search for octupole deformation in odd mass well-deformed light actinide nuclei, several degenerate parity doublets were predicted as the signature of intrinsic reflection asymmetry. Later, in experimental investigations of level schemes in many odd mass nuclei in this region, pairs of interleaved rotational bands with alternating positive and negative parity states and with the same set of level spins were found. The pair of sequences have the same K-value. The almost energy degenerate inter-pair of states with the same spin I but opposite parity are called parity-doublet states. The two pairs of sequences or bands should be simplex partners, like with $s = -i$ and the other with $s = +i$ corresponding to a single parity-mixed intrinsic orbital. A word of caution is required here: In the deformed shell model for reflection symmetric shapes, there could be closely spaced but unrelated Nilsson orbitals. Similar bands may arise from such orbitals.

Two pairs of yrast alternating parity bands (parity doublet bands) have been found in experimental investigations of the odd-N actinide nuclei ^{221}Ra [56], ^{223}Th [50, 57] and ^{225}Th [58].

Three pairs of parity doublet bands with $K^\pi = 1/2^\pm$, $K^\pi = 3/2^\pm$ and $K^\pi = 5/2^\pm$ have also been found in ^{223}Ra [13 and references therein].

Candidate parity doublet bands have also been found in some other odd-N nuclei in this mass region, like, ^{219}Ra [59–64], ^{219}Th [65] and ^{221}Th [51, 66].

Amongst the odd-Z nuclei, a mention could be made of ^{219}Ac [67–69], and ^{221}Ac [70], where candidate parity doublet bands have been observed.

As an example of parity-doublet bands, in Fig. 3.31 the level scheme of ^{223}Th [57] is shown. Out of the four yrast band sequences 1(a), 1(b), 2(a) and 2(b), bands 1(a)–1(b) and 2(a)–2(b) are the two alternating parity bands. The parity doublet states have same spin I opposite parity states $13/2^- - 13/2^+$, $17/2^- - 17/2^+$, ... $41/2^- - 41/2^+$, $45/2^- - 45/2^+$ in bands 1(a)–2(b) and $11/2^+ - 11/2^-$, $15/2^+ - 15/2^-$, ... $39/2^+ - 39/2^-$, $43/2^+ - 43/2^-$ states in 1(b)–2(a). The non-yrast bands 3(a) and 3(b) will not be discussed. For further details on these bands, see [57].

Several theoretical calculations have been done to interpret the parity doublet rotational bands observed in these odd mass nuclei. References to some of these works are [71–77, and references therein].

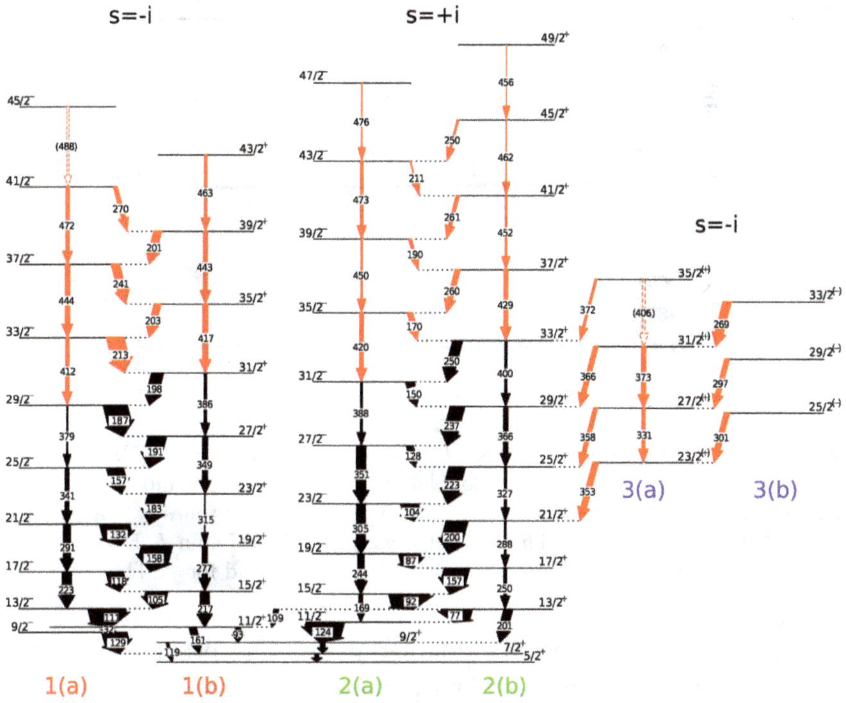

Fig. 3.31. The level scheme of ^{223}Th [57]. The transitions marked in black were already observed earlier in [50]. Figure is reproduced with permission from [57].

Figure 3.32 shows plots of the energy difference $\{E(I, s = +i) - E(I, s = -i)\}$ between the $\pi(+ \rightarrow -)$ and $\pi(- \rightarrow +)$ parity doublet states in the $s = +i$ and $s = -i$ yrast parity doublet bands respectively, in odd-N, $N = 133$ isotones ^{221}Ra and ^{223}Th and $N = 135$ ^{225}Th nuclei, as a function of spin I. Here, it is assumed that the energy separation between the $3/2^+$ ground state of the $s = -i$ band and the $5/2^+$ state of the $s = +i$ band in ^{225}Th is 31 keV [58], although this could not be firmly established experimentally. These parity doublet bands in these nuclei are almost degenerate in energy with minimum energy difference of 2.4 keV for the $15/2^-$, $15/2^+$ pair in ^{221}Ra and maximum energy difference of 67 keV for the $11/2^-$, $11/2^+$ pair in ^{225}Th. One of the general features of these plots is that the energy difference is staggering between $\Delta E[\pi(+ \rightarrow -)]$

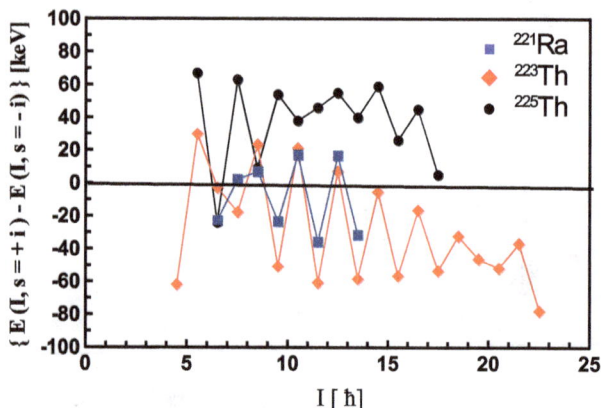

Fig. 3.32. The energy difference $\{E(I, s = +i) - E(I, s = -i)\}$ between the $\pi(+ \to -)$ and $\pi(- \to +)$ parity doublet states in the $s = +i$ and $s = -i$ yrast parity doublet bands respectively in odd-N, $N = 133$ isotones ^{221}Ra (■) and ^{223}Th (♦) and $N = 135$ ^{225}Th (●) nuclei, as a function of spin I. Experimental data for ^{221}Ra are taken from [56], for ^{223}Th from [57] and for ^{225}Th from [58].

and $\Delta E[\pi(- \to +)]$ parity doublet pairs. The other is that the staggering phase changes, that is, there is simplex inversion in all the parity doublet bands in these three nuclei. In ^{221}Ra, the simplex inversion takes place at spin $15/2^-$ or $17/2^+$ and then the energy difference staggering continues. In the ^{223}Th isotone, after the initial staggering at low spins, simplex inversion occurs at spin $13/2^+$ or $15/2^-$. Thereafter, the staggering is in phase with that in ^{221}Ra. A second simplex inversion occurs at a high spin of $39/2^-$ or $41/2^+$ in ^{223}Th. In the case of ^{225}Th nucleus, the simplex inversion is delayed, occuring at much higher spin of $23/2^-$ or $25/2^+$. The energy staggering is present before and after simplex inversion. At higher spins, the staggering is again in phase with that in the other two nuclei.

No theoretical explanation could be found in literature for this simplex inversion in these nuclei. Although a similar phenomena has been found in ^{145}Ba nucleus [78]. In [50], the staggering observed in the plot of $(E(I) - E(I-1))/(2I)$ as a function of I^2 for the positive and the negative parity bands in ^{223}Th (see their Fig. 11) has been discussed in terms of it being a combined effect of signature splitting between the bands and parity splitting.

3.8. Displacement Energy or Parity Splitting

As discussed earlier in Sec. 3.4, for even-even nuclei, the evolution of octupole correlations can be investigated through the behavior of displacement energy (parity splitting) as a function of spin. Comparison of displacement energy or parity splitting $\delta E(I)$ as a function of spin $(I - I_0)$ above the ground state spin I_0 is made in Fig. 3.33, for the $s = +i$ and $s = -i$, positive and negative parity yrast bands in the $N = 133$ isotones ^{221}Ra, ^{223}Th and $N = 135$ ^{225}Th nucleus with those in the neighboring even-even Ra and Th nuclei. A common general feature for the odd-N nuclei in these plots is that, as for the neighboring even-even nuclei, parity splitting $\delta E(I)$ decreases with increase in $(I - I_0)$. At a certain value of $(I - I_0)$ the parity splitting becomes zero, i.e., $\delta E(I) = 0$ and the nucleus can be said to acquire a stable octupole deformed shape. Another point to be noted is that at low spins, parity splitting in odd-N nuclei is smaller than that for the neighboring even-even nuclei. The $s = -i$ band in ^{225}Th, however, behaves somewhat differently. Here, the parity splitting is nearly equal to that in ^{224}Th neighbor but is still lower than in ^{226}Th. Also, a simplex dependence of parity splitting $\delta E(I)$ is observed for the $s = +i$ and $s = -i$ bands in the above odd-N nuclei. The $N = 133$ isotones ^{221}Ra and ^{223}Th, however, exhibit a similar increasing trend in simplex dependent parity splitting as a function of $(I - I_0)$.

The theoretical interpretation of the decrease of parity splitting with the increase in spin and the crossing of the $\delta E(I) = 0$ line is discussed in Sec. 3.4. The reduction of parity splitting observed experimentally as mentioned above at low spins in the odd-N nuclei as compared with those in the neighboring even-even nuclei, finds an explanation in the theoretical works [71, 79]. It has been shown in these calculations that there will be a reduction in parity splitting in

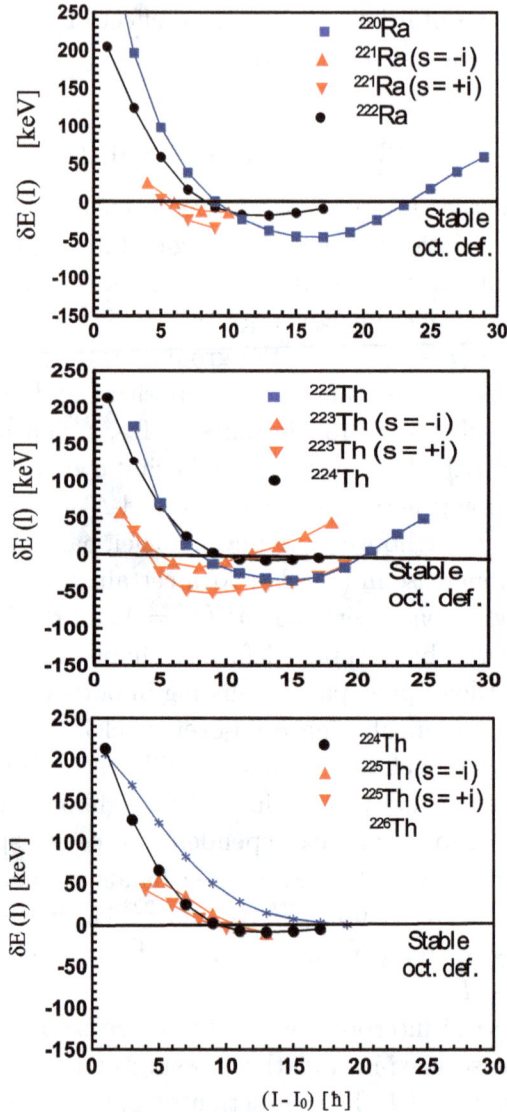

Fig. 3.33. Comparison of parity splitting $\delta E(I)$ for the $N = 133$ isotones ^{221}Ra, ^{223}Th and $N = 135$ ^{225}Th nucleus for the $s = -i$ and $s = +i$ bands with those in the neighboring even-even Ra and the Th isotopes respectively, as a function of spin $(I - I_0)$ above the ground state spin I_0. $I_0(^{221}$Ra$) = 5/2^+$, $I_0(^{223}$Th$) = 5/2^+$ and $I_0(^{225}$Th$) = 3/2^+$. $I_0 = 0$ for even-even nuclei. The experimental data for ^{220}Ra, ^{222}Ra, ^{222}Th, ^{224}Th and ^{226}Th nuclei are taken from [13 and references therein]. The experimental data for ^{221}Ra taken from [56], for ^{223}Th from [50, 57] and for ^{225}Th from [58].

odd-N nuclei as compared to even-even nuclei by a factor $\langle \pi \rangle$, the expectation value of the single particle parity operator.

3.9. Kinematic Moment of Inertia

Figure 3.34 shows the plot of kinematic moment of inertia as a function of rotational frequency $\hbar\omega$, for odd-N nuclei ^{223}Th (upper panel) [57] and ^{225}Th (lower panel) [58]. The moment of inertia in the two odd-N nuclei is larger compared to their even-even neighbors ^{222}Th (see Fig. 3.27, lower panel) and ^{224}Th (see Fig. 3.28, top right

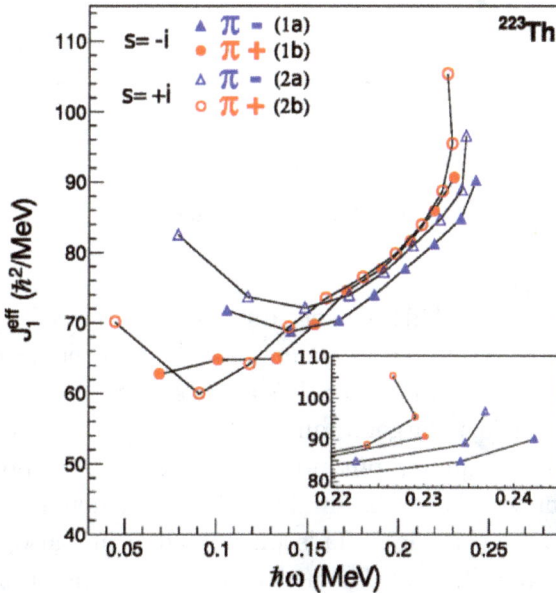

Fig. 3.34. Plots of kinematic moment of inertia as a function of rotational frequency for the positive and negative parity sequences with simplex quantum number $s = -i$ and $s = +i$ respectively, in ^{223}Th (upper panel) and ^{225}Th (lower panel). The insert (upper panel) shows a blow-up of the plot near the high-end of rotational frequency. Figures in upper and lower panels are reproduced with permission from [57] and [58] respectively. (**Note:** (i) The kinematic moment of inertia $\mathcal{J}^{(1)}$, in literature, has often been written as $\mathcal{J}_{\text{eff}}$ (or J_1^{eff}) when I_x, the angular momentum along the rotation axis comprises of both the collective rotation and the particle alignment contributions. (ii) In the lower panel, the units for $\mathcal{J}_{\text{eff}}$ should be $(\hbar^2\,\text{MeV}^{-1})$ and not $(\hbar\,\text{MeV}^{-1})$.).

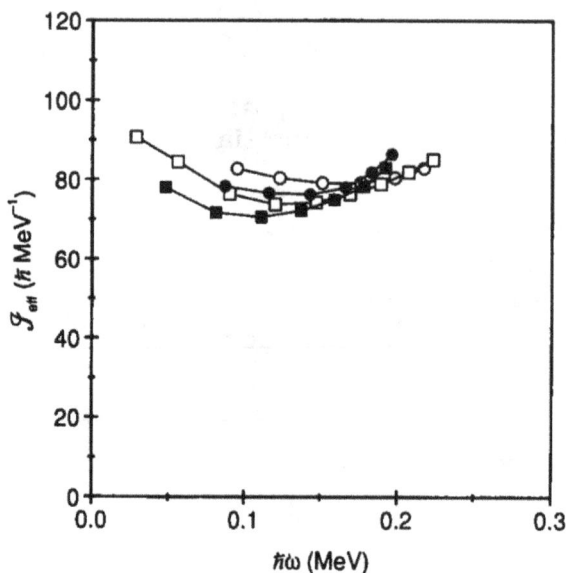

Fig. 3.34. (*Continued*)

panel), respectively. This is likely due to the unpaired nucleon — a neutron. At low rotational frequency, in ^{223}Th, in the negative parity bands for both the simplex bands $s = -i$ and $s = +i$, the kinematic moment of inertia is larger than that in the positive parity bands. In ^{225}Th, this is similar but not so spectacular. In both the odd nuclei, all the bands attain similar values of kinematic moment of inertia at $\hbar\omega \gtrsim 0.17\,\text{MeV}$. This phenomenon has also been found in the even-even Ra and Th nuclei, indicating the stabilization of octupole deformation. In ^{223}Th, in the region of the highest observed rotational frequencies, in the $s = +i$ bands, a backbend and an upbend are seen at $\sim 0.23\,\text{MeV}$ in the positive and the negative parity bands respectively. There is no such bending in the $s = -i$ bands. Both these phenomena can be seen in the insert in the figure for ^{223}Th. It may be mentioned here that a bandcrossing has been observed in the neighboring nucleus ^{222}Th within the same rotational frequency region (see Sec. 3.6).

3.10. Magnetic Dipole ($M1$) Transitions and Magnetic Dipole Moments

In the parity doublet bands observed in the odd mass nuclei in the actinide and the lanthanide regions, at low spins, in many rotational bands in these nuclei, strong low energy mostly magnetic dipole ($M1$) transitions with small E2 admixtures, are found. In this sub-section, we will discuss the level decay schemes in some of these nuclei and the properties of the M1 transitions. Also, where possible, the derived magnetic dipole moment of excited states will be compared with the experimentally measured ground state magnetic dipole moments.

3.10.1. *Theoretical framework*

In the strong coupling of the odd particle to the deformed nuclear core, the spin of the band head of the rotational band will be $I = K$, where K is the angular momentum projection on the symmetry axis. The M1 transitions between the members of the band are sensitive to the single particle components of the wavefunctions. For $K \neq 0$ bands, the M1 operator has components arising from the internal motion of the neutrons and the protons, characterized by the intrinsic gyromagnetic ratio, the g_K-factor and from the collective rotation of the nucleus as a whole, described by the g_R-factor [50]. Considering an even-even nucleus of charge Ze and mass A as a classical rotor and the constituent neutrons and protons contributing similarly to the rotation of the nucleus, the g_R-factor is given by [80]

$$g_R = J_p/(J_p + J_n) = Z/A \simeq 0.4 \qquad (3.6)$$

where J_p and J_n are the moments of inertia of the proton and the neutron groups, respectively. The experimental values of g_R are found to be smaller than Z/A [80]. The reasons for this discrepancy have been assigned as due to the different pairing forces of the protons and the neutrons, higher for the protons than for the neutrons. This may contribute to less deformed proton distribution than that for the neutrons. The contribution of neutrons to collective rotation would then be more as compared to that of the protons.

In an odd-A nucleus, the additional neutron or the proton contributes significantly to the moment of inertia J of the nucleus and therefore it follows from Eq. (3.6), that

$$g_R(\text{odd-}Z \text{ nucleus}) > g_R(\text{even-even nucleus})$$
$$g_R(\text{odd-}N \text{ nucleus}) < g_R(\text{even-even nucleus})$$

Experimentally, on an average, it is also found that the g_R-factors derived from measured magnetic moments of rotational bands in deformed odd-Z nuclei are larger than in the deformed odd-N nuclei [80].

The reduced transition probability $B(M1)$ for magnetic dipole ($M1$) transitions between rotational states in the strong coupling model can be expressed in terms of g_K, g_R, K and the spin I of the state from which the transition decays. For $K \neq 1/2$ and the $M1$ transition between $I_f = I_i - 1$ where I_i and I_f are the initial and the final spin of the states, respectively, the transition probability $B(M1)$ is given by the following expression [81]

$$B(M1) = \frac{3}{4\pi}\mu_N^2(g_K - g_R)^2 K^2 \frac{(I-K)(I+K)}{I(2I+1)} \qquad (3.7)$$

For a band with $K = 1/2$, the above expression for $B(M1)$ will contain one more term corresponding to an additional parameter b_o [82].

Measuring a lifetime of a level decaying by predominantly a $M1$ and a crossover $E2$ transition (see Fig. 3.35) together with experimental branching ratios will allow the determination of absolute $B(M1)$ and $B(E2)$ values. The $(g_K - g_R)$ value can be deduced using the determined B(M1) from the above expression.

The particle and the deformed core g-factors g_K and g_R can be related to the static magnetic dipole moment μ of a state of spin I in $K \neq 1/2$ band by the relation [82]

$$\mu = \frac{K^2}{I+1}(g_K - g_R) + I g_R \qquad (3.8)$$

The intrinsic structure of a strongly coupled or a deformation-aligned rotational band can also be inferred by measuring the

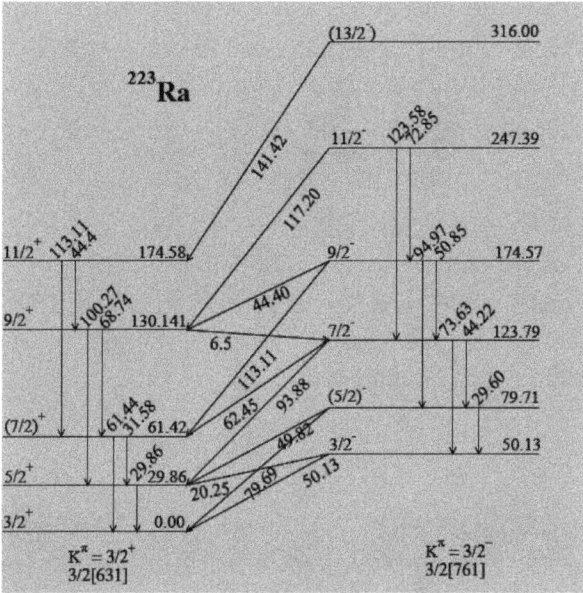

Fig. 3.35. Partial level scheme of ^{223}Ra in α-decay of ^{227}Th. Experimental data are taken from [13 and references therein].

intensity ratio of $\Delta I = 1(M1 + E2)$ and the $\Delta I = 2$ E2 gamma-ray transition branches from a level of spin I and the gamma-ray energies of the $(M1 + E2)$ and the $E2$ transitions decaying from it. The rotational model relation between these quantities, in the strong coupling limit, is given by [81]

$$\frac{\delta^2}{1+\delta^2} = \frac{2K^2(2I+1)}{(I+1)(I-1+K)(I-1-K)} \left(\frac{E_1}{E_2}\right)^5 \frac{I_r(\Delta I = 2)}{I_r(\Delta I = 1)} \quad (3.9)$$

where δ^2 is the $E2/M1$ mixing ratio, I is the spin of the state and K is the projection of angular momentum on the symmetry axis. The gamma-ray energies E_1 and E_2 for dipole and the quadrupole transitions respectively are in MeV.

The multipole mixing parameter δ for the mixed $(M1 + E2)$ transition can be deduced from the angular distribution or angular correlation measurements and the mixing ratio δ^2 can be determined from internal conversion data. The mixing ratio δ^2 can also be calculated from the above relation (3.9). Having obtained a good value of δ, the $(g_K - g_R)$ or the $(g_K - g_R)/Q_0$ value, can be determined for $K \neq 1/2$ band from the expression [83]

$$\delta = \frac{0.933 E_1 Q_0}{(g_K - g_R)\sqrt{(I^2 - 1)}} \qquad (3.10)$$

where E_1 is the transition energy in MeV and Q_0 is the intrinsic quadrupole moment.

For a state of spin I of a rotational band, the measured spectroscopic quadrupole moment Q_s is related to the intrinsic quadrupole moment Q_0 as in [82]

$$Q_s = \frac{3K^2 - I(I+1)}{(I+1)(2I+3)} Q_0 \qquad (3.11)$$

Relation (3.10) gives only the magnitude of the value of $(g_K - g_R)$. If the sign of the mixing parameter δ is inferred from angular distribution or the angular correlation measurements, then this phase is related to the sign of Q_0 and $(g_K - g_R)$ when $K \neq \frac{1}{2}$ as in [82]

$$\text{sign}\,\delta = \text{sign}\frac{g_K - g_R}{Q_0} \qquad (3.12)$$

The simplex partner bands (or parity doublet bands) in octupole deformed odd mass nuclei correspond to a single parity-mixed intrinsic orbital, these bands should be almost degenerate and exhibit similar magnetic dipole moments μ and the $(g_K - g_R)$ values, for the positive and the negative parity bands.

3.10.2. *Experimental determination of* $(g_K - g_R)/Q_0$ *and* $|g_K - g_R|$

In Fig. 3.35, the partial level scheme of ^{223}Ra nucleus depicting only the pair of $K^\pi = 3/2^\pm$ bands and the interconnecting gamma-ray transitions observed in α-decay of ^{227}Th, are shown [13 and references therein]. A number of inter band $E1$ transitions between the states of the negative and the positive parity bands and pairs of intra band $M1 + E2$ and $E2$ transitions decaying out from levels of the positive parity band and the levels of the negative parity band, were observed.

The parameter $(g_K - g_R)/Q_0$ was deduced in [84] for three $M1 + E2$ transitions each in the $K^\pi = 3/2^+$ and $K^\pi = 3/2^-$ bands in ^{223}Ra, namely, for the $5/2^+ \to 3/2^+$ 29.86 keV, $(7/2)^+ \to 5/2^+$ 31.58 keV and $9/2^+ \to (7/2)^+$ 68.74 keV transitions, and the $(5/2)^- \to 3/2^-$ 29.60 keV, $7/2^- \to (5/2)^-$ 44.22 keV and $9/2^- \to 7/2^-$ 50.85 keV transitions, respectively. The values of the $M1 + E2$ mixing parameter (δ) adopted for the above transitions were obtained from their [84] α-γ angular correlation measurements, branching ratios from available relative gamma-ray intensities for each pair of E2 and $M1 + E2$ transitions decaying from a level of spin I and the internal conversion data from literature. Table 3.2 from [84], gives the details of gamma-ray energies E_γ for the transitions of interest in the two bands, adopted values of the mixing parameter (δ), experimentally measured magnetic dipole moment (μ) values for $3/2^+$ ground state and $3/2^-$ state at 50.13 keV excitation energy [13 and references therein]. Also mentioned in the table are the deduced values of $-(g_K - g_R)/Q_0$ (column 4) and the average value of the rotational g-factor for both bands $g_R = 0.30 \pm 0.01$ which is smaller than $Z/A = 0.395$ for this nucleus as is the general trend for the experimental values of g_R [80]. The average value of g_R is obtained from g_R values from all measurements determined from $(g_K - g_R)/Q_0$ and the magnetic dipole moments (μ) for the bandhead states of the positive and the negative parity bands, and the deduced values of the

Table 3.2. Experimentally deduced values of $-\{(g_K - g_R)/Q_0\}$ in b^{-1} and g_K for $K^\pi = 3/2^\pm$ bands in ^{223}Ra, g_R (average value) $= 0.30 \pm 0.01$ for both bands [84].

$I \to (I-1)$	E_γ (keV)	δ	$-\{(g_K - g_R)/Q_0\}$	g_K
$K = 3/2^+$ band, $\mu = 0.2705(19)$ n.m.				
5/2	29.86	0.40(2)	0.030(2)	0.11(1)
7/2	31.58	0.27(1)	0.033(1)	0.10 (1)
9/2	68.72	0.46(4)	0.032(3)	0.10 (2)
$K = 3/2^-$ band, $\mu = +0.43(6)$ n.m.				
5/2	29.60	0.33(6)	0.037(7)	0.07(4)
7/2	44.10	0.55(8)	0.022(3)	0.16(2)
9/2	50.84	0.40(4)	0.027(3)	0.13(2)

intrinsic g-factors g_K (column 5). The intrinsic quadrupole moment
$Q_0 = +6.270 \pm 0.015\ b$ for $3/2^+$ ground state, was obtained (see
relation (3.11)) from the measured spectroscopic quadrupole moment
$Q_s = +1.254 \pm 0.003$ b [85 and references therein] in ^{223}Ra. It was
assumed that this value for the ground state quadrupole moment
is the same and remains constant in the low spin region for both
the negative and the positive parity bands. The deduced values of
$-(g_K - g_R)/Q_0$ for the positive parity and the negative parity states
in this $N = 135$ nucleus are plotted as a function of spin I in Fig. 3.36.
It is apparent from this plot that the parameter $(g_K - g_R)/Q_0$ is
similar and nearly constant for the two bands. This is a positive
argument in favor of the nucleus ^{223}Ra being reflection asymmetric
in the low spin regime.

In [84], theoretical values of g_K have also been calculated for
different values of the deformation parameters. For these calculations
"universal" Woods–Saxon parameters have been used. The small positive experimental values of g_K listed in column 5 in Table 3.2, for the
positive and the negative parity bands in ^{223}Ra, are compared with
the results of these theoretical calculations. The agreement between
theory and experiment requires deformed reflection asymmetry with
$\beta_3 \approx 0.05$. This result strengthens the conclusion arrived at from the
plot of $(g_K - g_R)/Q_0$ as a function of spin I, that at low spins, ^{223}Ra
is an octupole deformed nucleus.

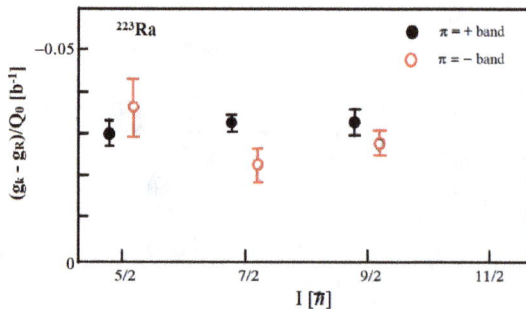

Fig. 3.36. Plot of the ratio $(g_K - g_R)/Q_0$ as a function of spin I for the intra
band M1 transitions in the positive and the negative parity states in the $K = 3/2$
bands in ^{223}Ra [84]. Figure is adopted from [84].

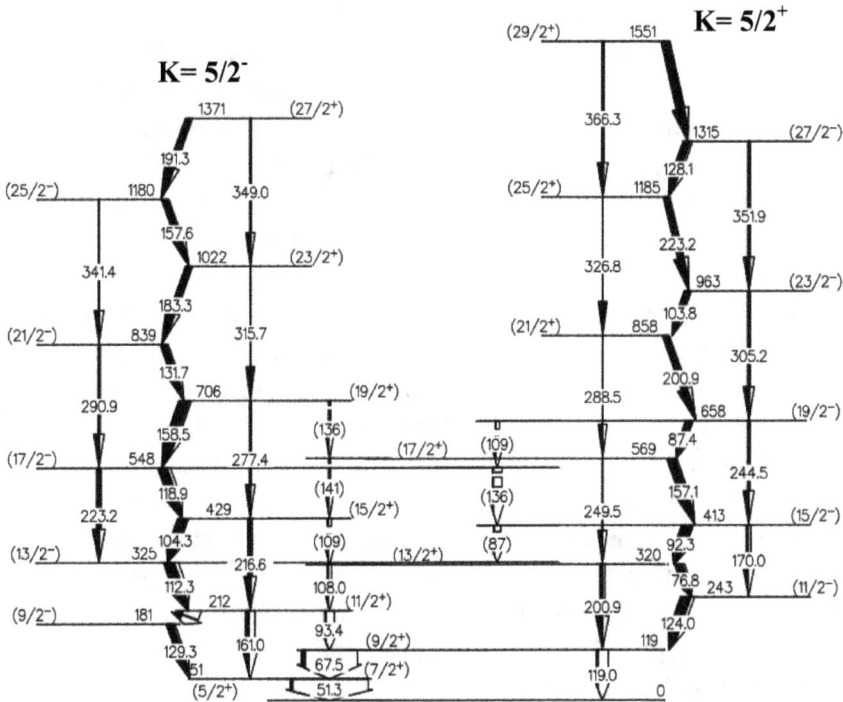

Fig. 3.37. Partial level scheme of ^{223}Th reproduced with permission from [86]. See also Fig. 3.31.

Figure 3.37 shows the level scheme of ^{223}Th [86]. Two pairs of $K^\pi = 5/2^\pm$ rotational bands each with strong intra band $E1$ and $E2$ transitions were observed. Here, one notices that there are states in each pair which have a counterpart in the other pair with same spin but opposite parity. Such pairs of states are the parity doublets. At low spins, strong low energy internally converted magnetic dipole $M1$ transitions which connect signature partners of the simplex bands, are observed, e.g., the 93.4 keV $(11/2^+) \rightarrow (9/2^+)$ transition between the positive parity bands and the 136 keV $(17/2^-) \rightarrow (15/2^-)$ transition between the negative parity bands. The 93.4 keV transition is mostly $M1$ with small (7%) $E2$ admixture and total internal conversion coefficient $\alpha_{\text{total}} = 5.7$.

The $|g_K - g_R|$ parameters for the 51.3 keV $(7/2^+) \rightarrow (5/2^+)$ $M1$ transition in the $K = 5/2^-$ band and the 67.5 keV $(9/2^+) \rightarrow (7/2^+)$ $M1$ transition in the $K = 5/2^+$ band were determined to be 0.39 ± 0.06 and $0.42 + 0.14$, -0.25, respectively from the measured $M1/E2$ mixing ratios δ^2 for the above transitions in ^{223}Th, as in [50]. The values for the $|g_K - g_R|$ parameter for the 93.4 keV $(11/2^+) \rightarrow (9/2^+)$ $M1$ transition in the $K = 5/2^-$ band and the 67.5 keV $(9/2^+) \rightarrow (7/2^+)$ and 87 keV $(15/2^=) \rightarrow (13/2^-)$ $M1$ transition in the $K = 5/2^+$ band were obtained from the measured values of the $B(M1)/B(E2)$ branching ratios [86]. All these experimentally obtained $|g_K - g_R|$ values are plotted as a function of spin I in Fig. 3.38. It is seen from this figure that within the error margin the values are similar for each $K = 5/2^{+ \text{ or } -}$ band and between the $K = 5/2^-$ and $K = 5/2^+$ bands. The similarity of the $|g_K - g_R|$ parameters is consistent with the interpretation that the nucleus ^{223}Th is reflection asymmetric. According to [87], the experimentally determined parameters do not agree with the theoretical values calculated for pure $\Omega = 5/2$ configurations of the states.

In Fig. 3.39, the level scheme of ^{225}Th as determined in [58], is shown. Two pairs of rotational bands, each band with intra band strong E1 and E2 transitions, have been observed. No inter band

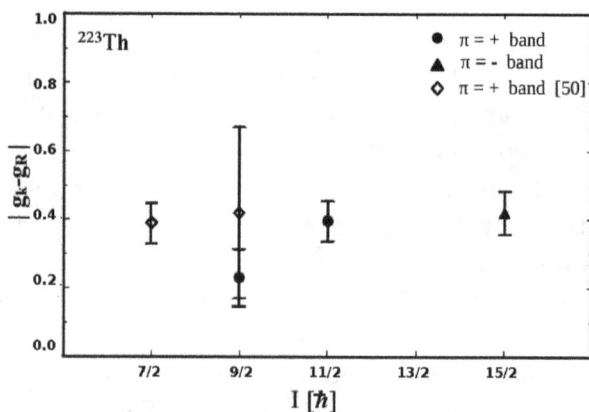

Fig. 3.38. Plot of the parameter $|g_K - g_R|$ as a function of spin I of the decaying level in ^{223}Th nucleus. The previous work is from [50]. Figure is adopted from [86].

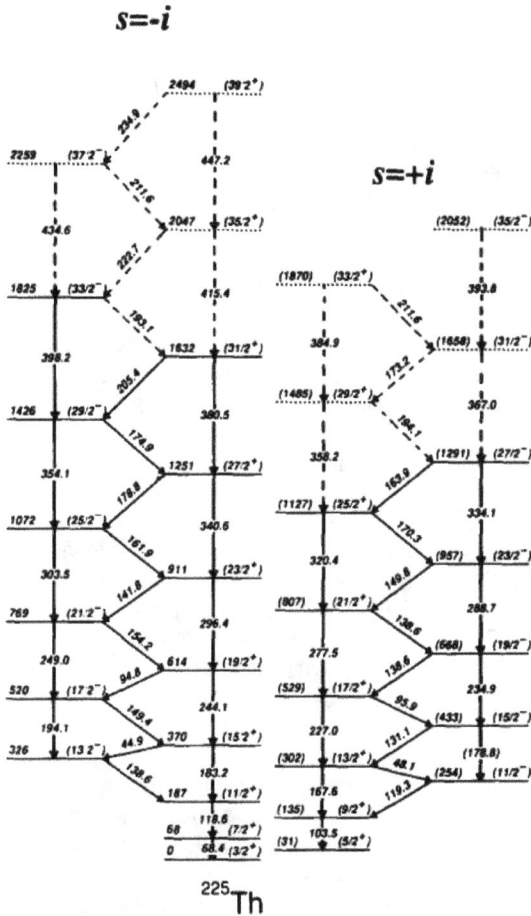

Fig. 3.39. Level scheme of ^{225}Th. Figure is reproduced with permission from [58].

connecting $M1$ transitions were seen. The authors have labeled the bands according to the simplex classification as the $K^\pi = 3/2^+$ and $K^\pi = 3/2^-$ bands. However, the mixing of the $s = +i$ band with the low lying $K = 5/2$ neutron state cannot be ruled out. Assuming that the observed bands are simplex partners, there are several parity doublets in the level scheme, pairs with the same spin and opposite parity.

References

1. D. Ward *et al.*, *Nucl. Phys. A* **406**, 591 (1983).
2. W. Bonin *et al.*, *Z. Phys. A, Atoms and Nuclei* **322**, 59 (1985).
3. P.D. Cottle, *Phys. Rev. C* **42**, 1264 (1990).
4. M.E. Debray *et al.*, *Nucl. Phys. A* **568**, 141 (1994).
5. M.E. Debray *et al.*, *Phys. Rev. C* **62**, 024304 (2000).
6. D. Bonatsos *et al.*, *Phys. Rev. C* **71**, 064309 (2005).
7. M.E. Debray *et al.*, *Phys. Rev. C* **73**, 024314 (2006).
8. M.E. Debray *et al.*, *Phys. Rev. C* **86**, 014326 (2012).
9. L.M. Robledo *et al.*, *Phys. Rev. C* **81**, 034315 (2010).
10. L.M. Robledo and P.A. Butler, *Phys. Rev. C* **88**, 051302(R). (2013).
11. K. Nomura *et al.*, *Phys. Rev. C* **89**, 024312 (2014) and references therein.
12. Z.P. Li *et al.*, *J. Phys. G: Nucl. Part. Phys.* **43**, 024005 (2016).
13. Brookhaven National Nuclear Data Center, ENSDF files:http://www.nndc.bnl.gov and references therein.
14. E. Parr *et al.*, *Phys. Rev. C* **94**, 014307 (2016).
15. N. Schulz *et al.*, *Phys. Rev. Lett.* **63**, 2645 (1989).
16. J.F. Smith *et al.*, *Phys. Rev. Lett.* **75**, 1050 (1995).
17. S. Raman *et al.*, *At. Data Nucl. Data Tables* **78**, 1 (2001).
18. Y.S. Chen and Z.C. Gao, *Phys. Rev. C* **63**, 014314 (2000).
19. J.F.C. Cocks *et al.*, *Nucl. Phys. A* **645**, 61 (1999).
20. W. Reviol *et al.*, *Phys. Rev. C* **74**, 044305 (2006).
21. W. Urban *et al.*, *Nucl. Phys. A* **613**, 107 (1997).
22. J.H. Hamilton *et al.*, *Acta Phys. Slovaca* **49**, 31 (1999).
23. W. Nazarewicz and P. Olanders, *Nucl. Phys. A* **441**, 420 (1985).
24. I. Ahmad and P.A. Butler, *Annu. Rev. Nucl. Part. Sci.* **43**, 71 (1993).
25. B. Ackermann *et al.*, *Nucl. Phys. A* **559**, 61 (1993).
26. R.V. Jolos and P. von Brentano, *Nucl. Phys. A* **587**, 377 (1995).
27. R.V. Jolos and P. von Brentano, *Phys. Rev. C* **60**, 064317 (1999).
28. R.V. Jolos, N. Minkov and W. Scheid, *Phys. Rev C* **72**, 064312 (2005).
29. R.V. Jolos and P. von Brentano, *Phys. Rev. C* **84**, 024312 (2011).
30. R.V. Jolos and P. von Brentano, *Phys. Rev. C* **49**, R2301 (1994).
31. S.J. Zhu *et al.*, *priv. comm.* (2012).
32. R.V. Jolos and P. von Brentano, *Phys. Rev. C* **92**, 044318 (2015).
33. W. Nazarewicz *et al.*, *Nucl. Phys. A* **467**, 437 (1987).
34. W. Urban *et al.*, *Eur. Phys. J A* **16**, 303 (2003).
35. P.A. Butler and W. Nazarewicz, *Rev. Mod. Phys.* **68**, 349 (1996).
36. J.M. Yao *et al.*, *Phys. Rev. C* **92**, 041304(R). (2015).
37. Y. Fu *et al.*, *Phys. Rev. C* **97**, 024338 (2018).
38. K. Nomura *et al.*, *Phys. Rev. C* **88**, 021303(R) (2013).
39. D. Bonatsos *et al.*, *Phys. Rev. C* **62**, 024301 (2000).
40. P.G. Bizzeti and A.M. Bizzeti-Sona, *Phys. Rev. C* **70**, 064319 (2004).
41. R. Bengtsson *et al.*, *At. Data Nucl. Data Tables* **35**, 15 (1986).
42. J. Simpson *et al.*, *J. Phys. (London)* **G10**, 383 (1984).
43. R. Holzmann *et al.*, *Phys. Rev. C* **31**, 421 (1985).

44. S.J. Zhu *et al.*, *Phys. Lett. B* **357**, 273 (1995).
45. P.A. Butler and L. Willmann, *Nucl. Phys. News*, **25**, 12 (2015).
46. P.A. Butler, *J. Phys. G: Nucl. Part. Phys.* **43**, 073002 (2016).
47. R. Bengtsson and S. Frauendorf, *Nucl. Phys. A* **327**, 139 (1979).
48. T. Rząca-Urban *et al.*, *Phys. Rev. C* **86**, 044324 (2012).
49. P. Bonche *et al.*, *Phys. Lett.* **175**, 387 (1986).
50. M. Dahlinger *et al.*, *Nucl. Phys. A* **484**, 337 (1988).
51. W. Reviol *et al.*, *Phys. Rev. C* **90**, 044318 (2014).
52. J.H. Hamilton *et al.*, *Acta Phys. Pol. B* **32**, 957 (2001).
53. S.P. Bvumbi *et al.*, *Phys. Rev. C* **87**, 044333 (2013).
54. W. Urban *et al.*, *Phys. Lett. B* **185**, 331 (1987).
55. R.R. Chasman, *Phys. Lett. B* **96**, 7 (1980).
56. J. Fernández - Niello *et al.*, *Nucl. Phys. A* **531**, 164 (1991).
57. G. Maquart *et al.*, *Phys. Rev. C* **95**, 034304 (2017).
58. J.R. Hughes *et al.*, *Nucl. Phys. A* **512**, 275 (1990).
59. T.C. Hensley *et al.*, *Phys. Rev. C* **96**, 034325 (2017).
60. P.D. Cottle *et al.*, *Phys. Rev. C* **33**, 1855 (1986).
61. P.D. Cottle *et al.*, *Phys. Rev. C* **36**, 2286 (1987).
62. M. Wieland *et al.*, *Phys. Rev. C* **45**, 1035 (1992).
63. L.A. Riley *et al.*, *Phys. Rev. C* **62**, 021301(R) (2000).
64. R.K. Sheline *et al.*, *Czech. J. Phys.* **51**, 111 (2001).
65. W. Reviol *et al.*, *Phys. Rev. C* **80**, 011304(R) (2009).
66. S.K. Tandel *et al.*, *Phys. Rev. C* **87**, 034319 (2013).
67. M.W. Drigert *et al.*, *Phys. Rev. C* **31**, 1977 (1985).
68. M.W. Drigert *et al.*, *Phys. Rev. C* **33**, 1344 (1986).
69. F. Cristancho *et al.*, *Phys. Rev. C* **49**, 663 (1994).
70. M. Aiche *et al.*, *Nucl. Phys. A* **567**, 685 (1994).
71. G.A. Leander and R.K. Sheline, *Nucl. Phys. A* **413**, 375 (1984).
72. G.A. Leander and Y.S. Chen, *Phys. Rev. C* **37**, 2744 (1988).
73. S. Ćwiok and W. Nazarewicz, *Nucl. Phys. A* **529**, 95 (1991).
74. G.G. Adamian *et al.*, *Phys. Rev. C* **70**, 064318 (2004).
75. X.T. He *et al.*, *Int. J. Mod. Phys. E* **15**, 1823 (2006).
76. N. Minkov *et al.*, *Phys. Rev. C* **76**, 034324 (2007).
77. N. Minkov, *Phys. Scr. T* **154**, 014017 (2013).
78. Y.-J. Chen *et al.*, *Phys. Rev. C* **91**, 014317 (2015).
79. D.M. Brink *et al.*, *J. Phys. G* **13**, 629 (1987).
80. E. Recknagel in *Nuclear Spectroscopy and Reactions*, Part C, ed. J. Cerny (Academic Press, N.Y., (1974)), p. 93.
81. P. Regan, *Post Graduate Experimental Techniques (4NET). Course Notes*, University of Surrey (2003).
82. K. Alder *et al.*, *Rev. Mod. Phys.* **28**, 432 (1956).
83. S. Mutto *et al.*, *Phys. Rev. C* **89**, 044309 (2014).
84. G.D. Jones *et al.*, *Eur. Phys. J. A* **2**, 129 (1998).
85. E. Browne, *Nuclear Data Sheets*, **65**, 669 (1992).
86. N. Amzal *et al.*, *Acta Phys. Pol.* **30**, 681 (1999).
87. P.A. Butler, *Phys. Scr. T* **88**, 7 (2000).

Chapter 4

High Spin Behavior of Pear-Shaped Nuclei-II

4.1. Introduction

In Chapter 3, the properties of pear-shaped nuclei were discussed which were derived from measuring the energy of high spin levels in such nuclei. In the present chapter, the properties of inter-band $\Delta I = 1$ $E1$ transitions are discussed that connect negative (positive) parity to positive (negative) parity neighboring levels between the negative and the positive parity bands and the intra-band $\Delta I = 2$ gamma-ray transitions between the same parity high spin levels ($E2$ transitions), in the above mentioned nuclei. We, also discuss $\Delta I = 3$ $E3$ transitions populated in Coulomb excitation. Such transitions in the decay mode in these nuclei are not observed because of the high multipolarity of the transitions. Section 4.2 is devoted to a discussion of the ratios of reduced electric dipole and quadrupole gamma-ray transition probability, $B(E1)/B(E2)$. In Section 4.3, absolute reduced electric dipole transition probabilities $B(E1)$ values are derived from the $B(E1)/B(E2)$ ratios and the absolute electric quadrupole transition probability $B(E2)$. In the absence of lifetime measurements of high spin levels in most cases, the latter quantity was derived from the $B(E2; 2^+ \rightarrow 0^+)$ measurements using a rigid rotor expression. This presents proof that the interconnecting $E1$ transitions in these octupole shaped nuclei are enhanced. Section 4.4 is devoted to a discussion on the intrinsic electric dipole moments D_o and their variation with nucleon number in Ra, Th, Ba isotopes and the $N = 88$ isotones. We also try to investigate why in ^{224}Ra and ^{146}Ba nuclei, highly depleted values of the dipole moments are observed. Section 4.5 pertains to the systematics of excitation energies of the 3^- states in the actinides and the lanthanides. This study probes further into the degree of octupole collectivity in nuclei. In the final section (Sec. 4.6), we consider experimental values and theoretical predictions of octupole transition probabilities $B(E3)$ and experimentally determined values of octupole moments Q_3. In ^{224}Ra, ^{226}Ra, ^{144}Ba and ^{146}Ba nuclei enhanced values of octupole moments were found in Coulomb excitation experiments. These results point to stable octupole or pear shape in these nuclei. A discussion regarding this aspect is also provided in this section.

4.2. $B(E1)/B(E2)$ Ratios

Absolute values of $B(E1)$, $B(E2)$ and also $B(M1)$ if $M1$ transitions are present in the decay of low-lying states, can be obtained from the lifetime measurements of excited states in nuclei and the experimental transition branching ratios. Lifetime measurements are done using the delayed coincidence, recoil distance or the Doppler shift attenuation methods. In these octupole deformed nuclei, there are experimental challenges in the population of these nuclei and for suitable experimental conditions in the lifetime measurements. As a result, only in very few cases, lifetime measurements are available. One has, therefore to depend on the $B(E1)/B(E2)$ ratios. This ratio, for the gamma-decay from a level with spin I is determined, from the gamma-ray branching ratio $I_\gamma(E1)/I_\gamma(E2)$ from that level for the $I \to (I-1)$ and $I \to (I-2)$ gamma-ray transitions respectively, under the assumption that the positive parity ground-state band and the low negative parity band have both pure $K = 0$, using the following relation [1]:

$$\frac{B(E1)}{B(E2)} = \left(\frac{E_\gamma(E2)}{E_\gamma(E1)}\right)^3 \frac{(E_\gamma(E2))^2}{1.30 \times 10^6} \frac{I_\gamma(E1)}{I_\gamma(E2)} \qquad (4.1)$$

where $E_\gamma(E1)$ and $E_\gamma(E2)$ are the gamma-ray energies in MeV and the $B(E\lambda)$'s in $e^2 \cdot \text{fm}^{2\lambda}$. In this manner, the $B(E1)/B(E2)$

ratios for a large number of levels in the light actinide and lanthanide nuclei have been determined. Further, to obtain the $B(E1)$ value from $B(E1)/B(E2)$ ratio, one needs to know the of experimental value of $B(E2)$ as a function of spin I which is not available for these nuclei. One alternative is, in an even-even nucleus, obtain the $B(E2)$ value from the measured lifetime, if available, for the 2^+ state and consider that it [i.e., $B(E2)$] remains constant as a function of spin in a nucleus. Although, this may not be strictly true.

In ^{218}Ra ($Z = 88$, $N = 130$), lifetime measurements have been done [2] for some low spin states by the Doppler shift recoil distance method. The excited states in this nucleus were populated by the inverse reaction ^{13}C(^{208}Pb, 3n) to obtain large recoil velocities ($\beta \approx 10\%$). From the lifetime measurements, absolute $B(E1)$ and $B(E2)$ values were obtained as a function of initial spin. Due to experimental difficulties, the accuracy of measurement was limited. However, a few data points with large error bars did not facilitate a definite conclusion about the constancy of $B(E2)$ values with spin. As for the $B(E1)$ values, there is a rise in the value from $B(E1) \approx 2 \times 10^{-3}$ to 6×10^{-3} W.u., between $I = 6$ to 8.

We shall now discuss the systematics of $B(E1)/B(E2)$ ratios as a function of initial spin I for a number of even-even and odd-N Ra and Th nuclei. In Fig. 4.1 the plots of the ratios versus initial spin I for ^{218,220,222}Ra isotopes, are shown. In Fig. 4.2, similar plots are shown, for ^{224,226,228}Ra isotopes.

The level scheme of ^{218}Ra ($N = 130$) [3 and references therein] consists mainly of a positive parity ground state band and a negative parity octupole band. Both the bands are connected by fast $E1$ transitions. The energy levels of these bands exhibit nearly equidistant spacings. The $E(4^+)/E(2^+)$ energy ratio for this nucleus is 1.90. This nucleus, therefore, can be classified as a quadrupole vibrational nucleus. The $B(E1)/B(E2)$ ratios for this nucleus show a decreasing trend as a function of increasing spin (Fig. 4.1, top panel). Further, no general difference can be inferred for the $B(E1)/B(E2)$ ratios for decay from either the positive parity or the negative parity states. The decreasing trend in $B(E1)/B(E2)$ ratios has been explained in [4] as due to the crossing of the negative parity octupole

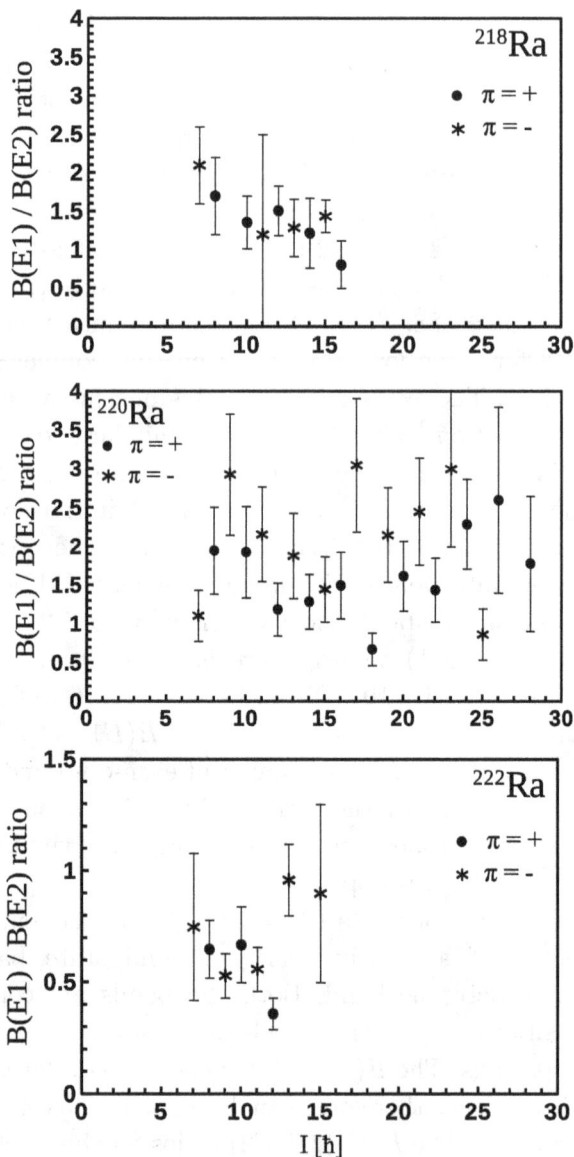

Fig. 4.1. Plots of $B(E1)/B(E2)$ ratios in units of $[10^{-6}\,\text{fm}^{-2}]$ versus initial spin I, for even-even ^{218,220,222}Ra isotopes. The ratios for decay from the +ve and the −ve parity states are marked by filled circle (•) and star (∗), respectively. Experimental data for ^{218}Ra and ^{220}Ra are taken from [3 and references therein] and for ^{222}Ra from [7].

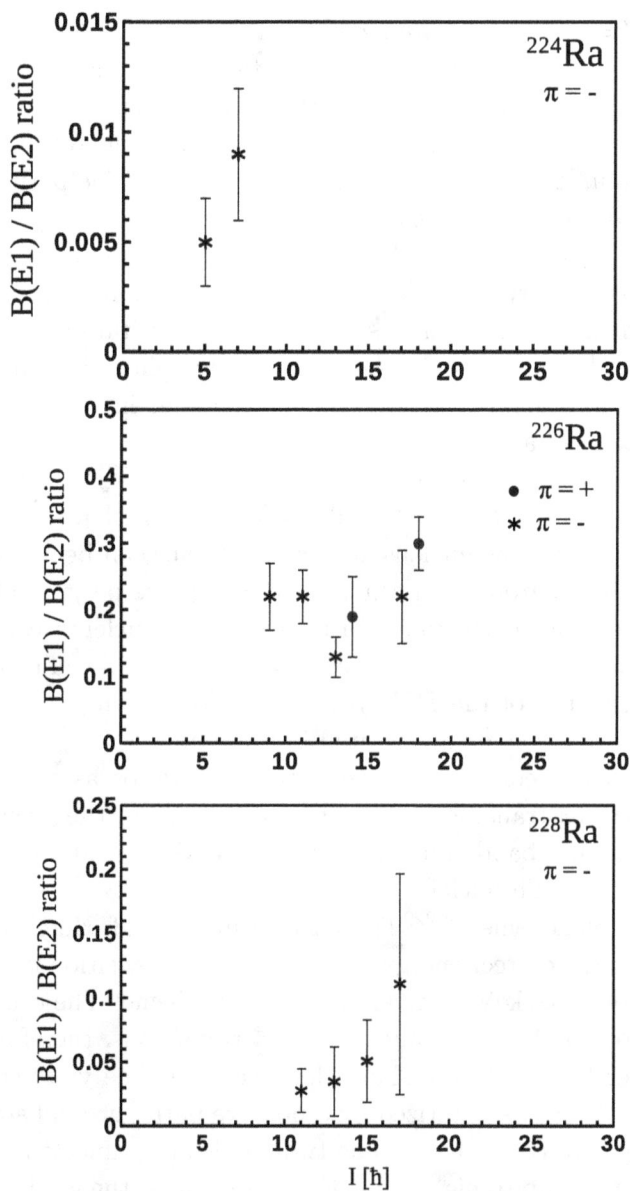

Fig. 4.2. Plots of $B(E1)/B(E2)$ ratios in units of $[10^{-6}\,\text{fm}^{-2}]$ versus initial spin I, for even-even ^{224,226,228}Ra isotopes. The ratios for decay from the +ve and the −ve parity states are marked by filled circle (•) and star (∗), respectively. Experimental data for ^{224,226,228}Ra nuclei are taken from [7].

band at $I = 11^-$ by the even spin negative parity two quasi-particle neutron band [5]. The mixing of these two bands may be the reason for the reduction of the $E1$ strength at high spins.

The $N = 132$ ^{220}Ra nucleus has been investigated [6] to very high spins up to 28^+ (possibly to 30^+) in the positive parity ground state band and 29^- (possibly to 31^-) in the negative parity band (see Fig. 3.25 for level scheme). Apart from the intra-band $E2$ transitions, the two bands are connected by inter-band fast $E1$ transitions up to the highest spins. The level excitation energy ratio $E(4^+)/E(2^+) = 2.29$ indicates that at low spins, this nucleus has rather a vibrational character but as the spin increases, the level scheme exhibits a rotational pattern.

In Fig. 4.1 (middle panel) for the nucleus ^{220}Ra, no clear-cut dependence of the $B(E1)/B(E2)$ ratios as a function of initial spin I is observed. In this nucleus also, no difference can be made for the ratios for decay from the positive or the negative parity states.

Plots similar to the above for the even-even nuclei ^{222}Ra (Fig. 4.1 bottom panel) and ^{224,226,228}Ra (Fig. 4.2), also do not show any spin dependence of the $B(E1)/B(E2)$ ratio, although the available number of data points are limited in these nuclei.

Let us now consider the $B(E1)/B(E2)$ ratios as a function of initial spin for transitions decaying from levels in the ground state positive parity band and the negative parity band in even-even ^{220,222,224,226,228}Th nuclei.

The level scheme of ^{220}Th is shown in Fig. 4.3. Here, it may be noted that the correct energy for the $2^+ \rightarrow 0^+$ transition is 386.5 keV [8] and not 373.3 keV as given in the level scheme. The order of the energies of the $4^+ \rightarrow 2^+$ and $2^+ \rightarrow 0^+$ transitions, therefore, needs to be interchanged in the shown level scheme. The yrast structures (Bands (a)) are characterized by a positive parity ground state band and a negative parity band. The two bands are connected by strong $E1$ transitions up to high spins. For this nucleus, the level excitation energy ratio $E(4^+)/E(2^+) = 1.97$ indicates that at low spins, it has a normal vibrational character.

In Fig. 4.4 (upper panel), the $B(E1)/B(E2)$ ratios are shown as a function of initial spin I for the $N = 130$ ^{220}Th nucleus. Our

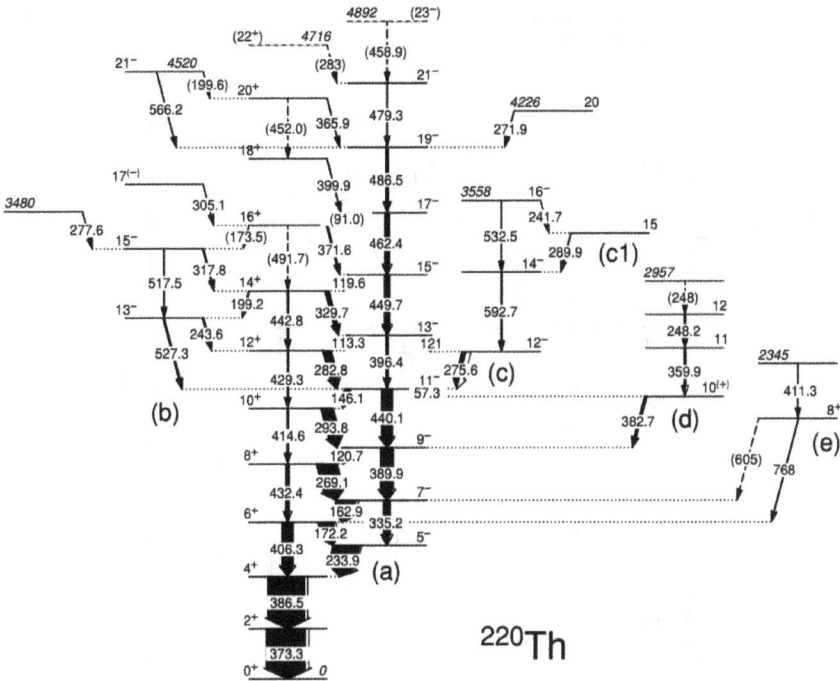

Fig. 4.3. Level scheme of ^{220}Th [9]. The order of the energies of the $4^+ \rightarrow 2^+$ and $2^+ \rightarrow 0^+$ transitions needs to be interchanged. The correct energy for the $2^+ \rightarrow 0^+$ transition is 386.5 keV [8] and not 373.3 keV as given in the level scheme. See [3] for level energies. Figure is reproduced with permission from [9].

discussion of these ratios for this nucleus is based on [9]. In the spin range 6 to $\sim$12, the ratios for the decay from the positive parity states are higher in comparison to those from the negative parity states. This can be explained in terms of energy staggering between the $E1$ transitions originating from the positive parity levels and the negative parity levels. The $E1$ transitions from the positive parity levels are favored and they compete with the branched $E2$ transitions. The high value of the $B(E1)/B(E2)$ ratio at the yrast state at spin 13^- is due to the rotation alignment effect (see Fig. 3.24) at which there is a loss of the $E2$ strength. The low value of the ratio at spin 14^+ is due to the division of the $E1$ strength at this level. In addition to the 329.7 keV $E1$ transition from the 14^+ level

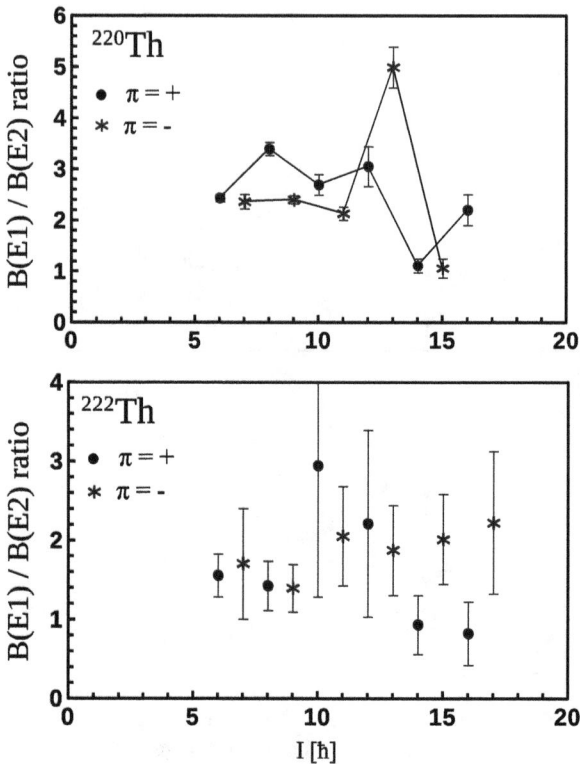

Fig. 4.4. Plots of $B(E1)/B(E2)$ ratios in units of $[10^{-6}\,\mathrm{fm}^{-2}]$ versus initial spin I, for even-even ^{220,222}Th isotopes. The ratios for decay from the +ve and the −ve parity states are marked by filled circle (•) and star (*), respectively. Experimental data for ^{220}Th are taken from [9] and for ^{222}Th from [6, 12].

to the 13^- 2555.2 keV yrast level, there is another $E1$ transition of 199.2 keV decaying out of the same level to nearby 13^- yrare state at 2685.8 keV. (The level energies of the two 13^- states are not mentioned on the level scheme.).

The $B(E1)/B(E2)$ ratios for $N = 132$ ^{222}Th nucleus are also shown as a function of initial spin I in Fig. 4.4 (lower panel). This plot does not exhibit any angular momentum dependence of the ratios. Also, no difference was considered for the ratios of decay from the positive or the negative parity states. For this nucleus, the level excitation energy ratio $E(4^+)/E(2^+) = 2.40$ and $\beta = 0.153$ [10], that indicates that it has a normal transitional type character at low spins.

Let us now consider the $N = 134$ ^{224}Th nucleus. The $B(E1)/B(E2)$ ratios for this nucleus are shown as a function of initial spin I in Fig. 4.5 (top panel). From this plot, an increase in the value of this ratio with increasing spin is observed. Whether this change is due to the variation of $B(E2)$ with spin or due to the variation of $B(E1)$ can only be known if the absolute values of $B(E1)$ and $B(E2)$ are available from level lifetime measurements.

In ^{226}Th, $B(E1)/B(E2)$ ratios as a function of spin (see Fig. 4.5, middle panel) remain almost constant in the spin range $I = 9 - 19$ at a level of 0.47×10^{-6} fm^{-2}. No comment on this ratio can be made in ^{228}Th nucleus because of the paucity of data points as a function of spin (see the bottom panel of the figure).

Below, the $B(E1)/B(E2)$ ratios as a function of initial spin I minus g.s. spin I_o, $(I - I_o)$ are discussed for the odd-N ^{219}Ra and ^{221,223,225}Th nuclei.

In Fig. 4.6, a plot of the $B(E1)/B(E2)$ ratios as a function of $(I - I_o)$ is shown for the $K^\pi = 1/2^+$ and $K^\pi = 1/2^-$ ground state bands in the odd-N ($N = 131$) ^{219}Ra nucleus. These bands have been interpreted in terms of weak coupling of the $g_{9/2}$ neutron to the ^{218}Ra soft-quadrupole core. No spin dependence is seen for the ratios. The ratio maintains at a level of 1.7×10^{-6} fm^{-2} [11].

In Fig. 4.7, the $B(E1)/B(E2)$ ratios as a function of $(I - I_o)$ are plotted for the $s = -i$ and the $s = +i$ bands in the odd-N ^{221,223,225}Th isotopes. No systematic spin dependence is found in any of these isotopes. However, in ^{221}Th, for the high spin range $(I - I_o) = 12$ to 16, a drop in the $B(E1)/B(E2)$ ratios is seen with the increase in spin.

In Figs. 4.8, the $B(E1)/B(E2)$ ratios versus spin are shown for the $N = 88$ ^{144}Ba, ^{146}Ce and ^{148}Nd isotones. Within the scatter of the data points, from $I = 8 - 16$, the $B(E1)/B(E2)$ ratios do not exhibit any perceptible spin dependence for ^{144}Ba nucleus. For the ^{146}Ce and ^{148}Nd isotones, the number of data points are too few to discern

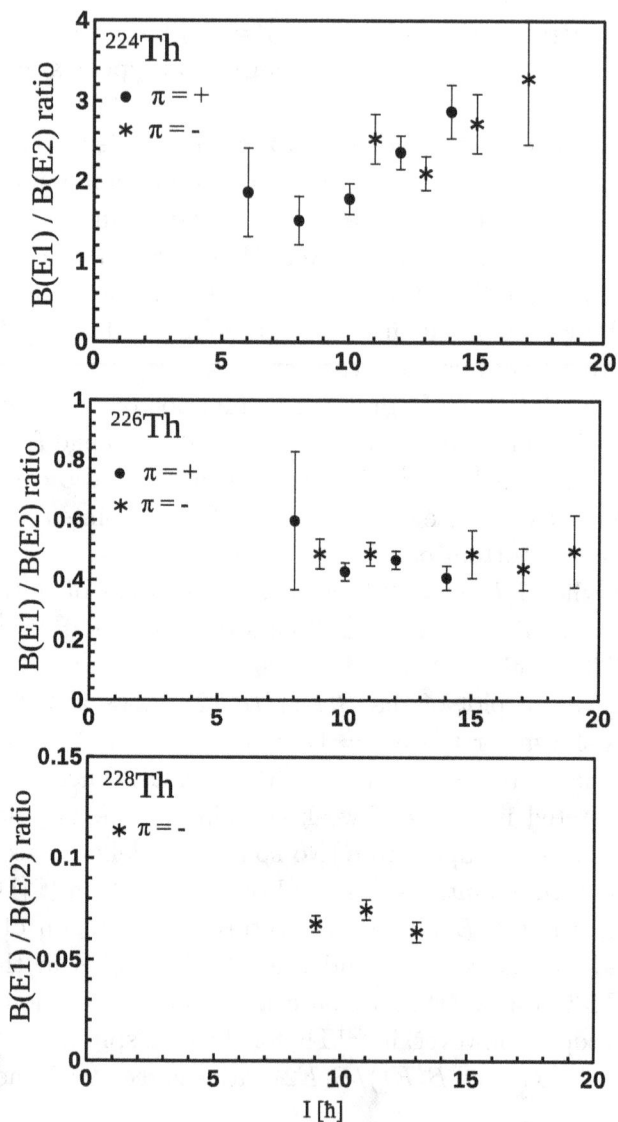

Fig. 4.5. Plots of $B(E1)/B(E2)$ ratios in units of $[10^{-6}\,\mathrm{fm}^{-2}]$ versus initial spin I, for even-even ^{224,226,228}Th isotopes. The ratios for decay from the +ve and the −ve parity states are marked by filled circle (•) and star (∗), respectively. Experimental data for ^{224,226,228}Th are taken from [13].

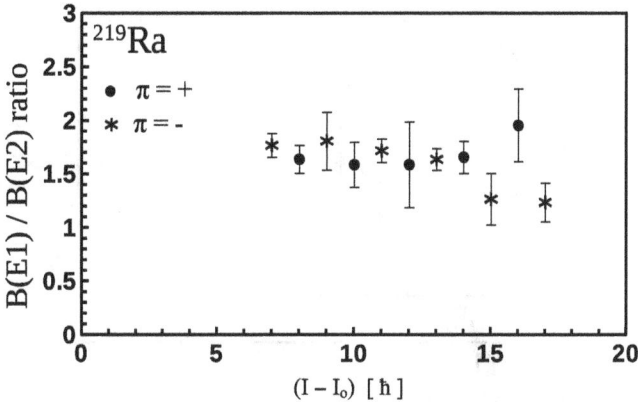

Fig. 4.6. Plot of $B(E1)/B(E2)$ ratios in units of $[10^{-6}\,\text{fm}^{-2}]$ versus initial spin I minus g.s. spin I_0, $(I - I_0)$ for the $K^\pi = 1/2^+$ and $K^\pi = 1/2^-$ bands in the odd-N $(N = 131)$ nucleus ^{219}Ra. The ratios for decay from the +ve and the −ve parity states are marked by filled circle ($\bullet$) and star ($*$), respectively. Experimental data for this nucleus are taken from [3 and references therein, [11]]. I_0 (g.s. ^{219}Ra) = $(7/2)^+$ from [3].

any spin dependence. Figure 4.9 [14] is a plot of the $B(E1)/B(E2)$ ratios versus spin for the other $N = 88$ transitional nucleus ^{150}Sm. The ratio levels off at $\sim 0.7 \times 10^{-6}\,\text{fm}^{-2}$, after spin around $I = 14$ at which the delayed alignment of a $i_{13/2}$ pair of neutrons occur in the positive parity band.

In [15], a theoretical attempt has been made to investigate angular momentum dependent phase transition to octupole deformation in the above $N = 88$ isotones. These calculations were done by extending the angular momentum constrained Hartree–Fock–Bogoliubov theory with Gogny force. Amongst other energy level properties, the $B(E1)/B(E2)$ ratios were calculated as a function of spin and compared with the available experimental data (see their Fig. 5). The agreement between theory and experiment was found to be good for ^{144}Ba, ^{146}Ce and ^{148}Nd isotones. However, for the ^{150}Sm isotone, the theoretical $B(E1)/B(E2)$ ratios at high spins were much higher than the experimental values, may be due to shape fluctuations in the spin region in this transitional nucleus.

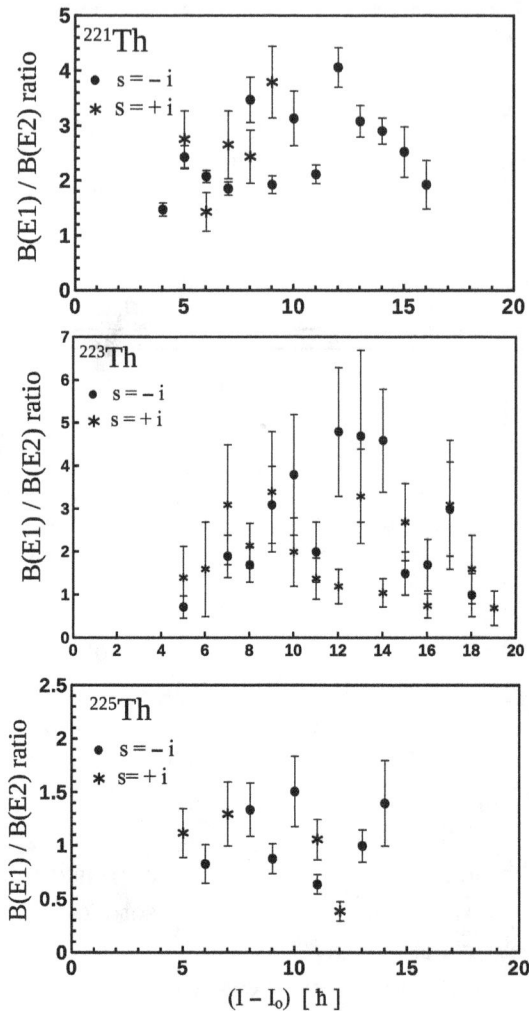

Fig. 4.7. Plots of $B(E1)/B(E2)$ ratios in units of $[10^{-6}\,\mathrm{fm}^{-2}]$ for the odd - N ^{221,223,225}Th isotopes versus initial spin I minus g.s. spin I_o, $(I-I_o)$. $I_o = (7/2^+)$, $(5/2)^+$, $(3/2^+)$ for the ground states of ^{221}Th, ^{223}Th and ^{225}Th respectively. The ratios for the $s = -i$ bands are marked by ($\bullet$) and for the $s = +i$ bands by ($*$). No symbol distinction has been made for the data points of the decaying states with +ve and −ve parities, within the same s band. Experimental data for ^{221}Th is taken from [16], for ^{223}Th from [17] and for ^{225}Th from [1]. It may be mentioned here that in ^{225}Th, the gamma-ray intensity values in $s = -i$ and $s = +i$ bands used are those with 118 keV and 103 keV L2 internal conversion electron lines as coincidence gates respectively, from [1].

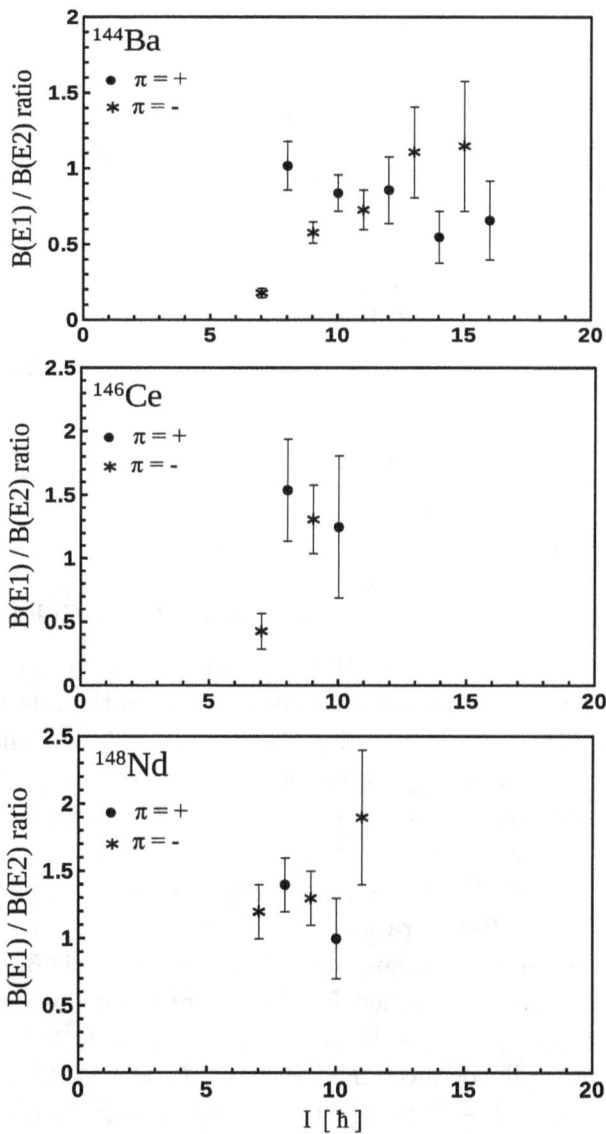

Fig. 4.8. Plots of $B(E1)/B(E2)$ ratios in units of $[10^{-6}\,\mathrm{fm}^{-2}]$ versus initial spin I, for the $N = 88$ ^{144}Ba, ^{146}Ce and ^{148}Nd isotones of decay from the positive parity states of the ground state band and from the states of the negative parity band. The ratios of decay from the +ve and the −ve parity states are marked by filled circle (•) and star (∗), respectively. Experimental data for ^{144}Ba, ^{146}Ce and ^{148}Nd nuclei are from [18–20] respectively.

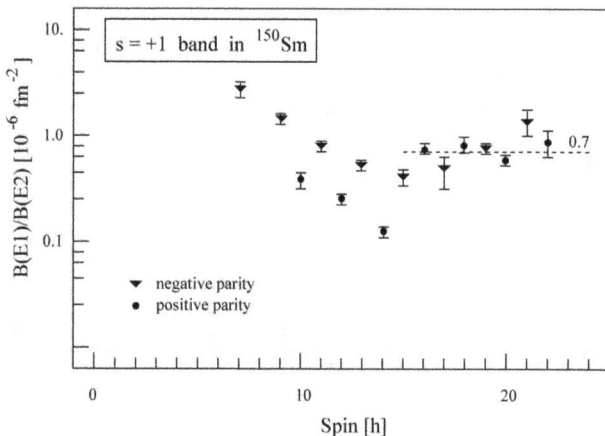

Fig. 4.9. Plot of $B(E1)/B(E2)$ ratios versus initial spin, for the $N = 88$ isotone ^{150}Sm. This figure is reproduced with permission from [14].

4.3. Electric Dipole Transition Probabilities $B(E1)$

Having discussed the $B(E1)/B(E2)$ ratios for the nuclei exhibiting octupole correlations in the actinide and the lanthanide nuclei, let us now pay attention to obtaining the absolute values of the reduced electric dipole gamma-ray transition probability $B(E1)$. As already mentioned earlier, the absolute values of $B(E1)$ can be obtained from level lifetime measurements but such measurements are mostly not available, one, therefore, has to depend on the experimentally measured $B(E1)/B(E2)$ ratios.

In an even-even nucleus, the absolute value of $B(E1; I_i \rightarrow I_f = I_i - 1)$ can be obtained from experimentally measured value of $B(E1; I_i \rightarrow I_f = I_i - 1)/B(E2; I_i \rightarrow I_f = I_i - 2)$ ratio and the experimentally deduced $B(E2; I_i \rightarrow I_f = I_i - 2)$ value. The $B(E2; I_i \rightarrow I_f = I_i - 2)$ value with respect to that for the $2^+ \rightarrow 0^+$ transition in a rotational band with $K = 0$, is obtained using the following formula [21]:

$$B(E2; \ I_i \rightarrow I_f = I_i - 2) = \frac{5}{16\pi} \langle I_i 020 | (I_i - 2)0 \rangle^2 e^2 Q_0^2 \qquad (4.2)$$

and

$$e^2 Q_0^2 = \frac{16\pi}{5} B(E2; \; 0^+ \to 2^+) \tag{4.3}$$

$$= 16\pi B(E2; \; 2^+ \to 0^+) \tag{4.4}$$

since $B(E2; \; 2^+ \to 0^+)$ i.e., $B(E2)\downarrow$ is related to $B(E2; \; 0^+ \to 2^+)$ i.e., $B(E2)\uparrow$ by the relation

$$B(E2)\downarrow = [(2I_f + 1)/(2I_i + 1)]B(E2)\uparrow \tag{4.5}$$

with $I_f = 0$ and $I_i = 2$, this reduces to

$$B(E2)\downarrow = [1/5]B(E2)\uparrow \tag{4.6}$$

Therefore, from Eqs. (4.2) and (4.4)

$$B(E2; \; I_i \to I_f = I_i - 2)$$
$$= 5\langle I_i 020|(I_i - 2)0\rangle^2 B(E2; \; 2^+ \to 0^+) \tag{4.7}$$

$$= 5\frac{3I_i(I_i - 1)}{2(2I_i - 1)(2I_i + 1)}B(E2; \; 2^+ \to 0^+) \tag{4.8}$$

using

$$\langle I_i 020|(I_i - 2)0\rangle^2 = \frac{3I_i(I_i - 1)}{2(2I_i - 1)(2I_i + 1)} \tag{4.9}$$

It may be useful to remind here that the relations (4.7)–(4.9), can be applied only in the rotational limit.

The adopted (recommended) $B(E2)\uparrow$ values in even-even nuclei with $Z = 2 - 104$ are tabulated in [22].

The $B(E1; \; I_i \to I_f = I_i - 1)$ i.e., $B(E1)$ can be obtained from

$$B(E1; \; I_i \to I_f = I_i - 1)$$
$$= \frac{B(E1; \; I_i \to I_f = I_i - 1)}{B(E2; \; I_i \to I_f = I_i - 2)} \times B(E2; \; I_i \to I_f = I_i - 2) \tag{4.10}$$

The reduced transition probability $B(E1)$ obtained in the above manner can be expressed in units of the Weisskopf estimate [23] of single particle reduced transition probability $B(E1)_w$. It is to be mentioned here that the Weisskopf estimate is not to be considered for reproducing the experimental data but to be used as a scale. The numerical value of $B(E1)_w = 0.06446\,A^{2/3}\,e^2.\mathrm{fm}^2$ [24, 25], where A is the mass number. The $B(E1)/B(E1)_w$ ratio provides a useful way for comparing the reduced electric dipole gamma-ray transition probabilities in different nuclei. In Table 4.1, are listed values of $B(E1)/B(E1)_w$ obtained at the mentioned spin in each nucleus using the above mentioned procedure, in the even-even Ra and Th isotopes. As compared to these ratios in the Ra and Th isotopes, typical values of the $B(E1)/B(E1)_w$ ratios are below 10^{-5} in nuclei which do not exhibit octupole correlation behavior (for details see [26]). This shows that the $E1$ transitions between the opposite parity

Table 4.1. The values of $B(E1)/B(E1)_w$ ratio obtained at the mentioned spins in the even-even Ra and Th nuclei, are listed below. The experimental $B(E1)/B(E2)$ ratios in these nuclei were deduced as mentioned in detail in Sec. 4.2 (see also Figs. 4.1, 4.2, 4.4 and 4.5). The experimental $B(E2;\ 0^+ \to 2^+)$ i.e., $B(E2)\uparrow$ values in the below mentioned nuclei were taken from [22]: $B(E2;\ 0^+ \to 2^+)$ for $^{218}\mathrm{Ra} = 1.00^{+0.10}_{-0.09},\ ^{222}\mathrm{Ra} = 4.51 \pm 0.36,$ $^{224}\mathrm{Ra} = 3.990 \pm 0.052,\ ^{226}\mathrm{Ra} = 5.16 \pm 0.13,\ ^{228}\mathrm{Ra} = 5.98 \pm 0.20,$ $^{222}\mathrm{Th} = 2.98 \pm 0.25,\ ^{224}\mathrm{Th} = 3.96^{+0.29}_{-0.25},\ ^{226}\mathrm{Th} = 6.82 \pm 0.35,\ ^{228}\mathrm{Th} = 7.05 \pm 0.12$ in $e^2 b^2$ units.

Nucleus	Spin [ℏ]	$B(E1)/B(E1)_w^{\dagger a)}$
^{218}Ra	12	$2.20(50) \times 10^{-3}$
^{222}Ra	12	$2.36(49) \times 10^{-3}$
^{224}Ra	7	$\mathbf{4.9(1.6) \times 10^{-5}}$
^{226}Ra	11	$1.62(29) \times 10^{-3}$
^{228}Ra	11	$2.4(1.4) \times 10^{-4}$
^{222}Th	9	$5.86(1.34) \times 10^{-3}$ [b)]
^{224}Th	10	$10.0(1.2) \times 10^{-3}$ [b)]
^{226}Th	10	$4.13(36) \times 10^{-3}$ [b)]
^{228}Th	11	$7.51(51) \times 10^{-4}$ [b)]

[a)] $B(E1)_w = 0.06446\,A^{2/3}\,e^2\,\mathrm{fm}^2$.
[b)] See also [27].

levels of the positive and the negative parity bands in these nuclei are fast or enhanced. The case of ^{224}Ra is an exception where the $B(E1)/B(E1)_w$ ratios are small. The ratio is also available at spin 5 $\hbar$ (not mentioned in the table). This quenching of the $E1$ strength in this nucleus is discussed later in Sec. 4.4.

In Table 4.2 are listed the values of $B(E1)/B(E1)_w$ ratio of decay from different spin states in the even-even ^{144}Ba, ^{146}Ce and ^{148}Nd

Table 4.2. The values of $B(E1)/B(E1)_w$ obtained at different spins in the even-even $N = 88$ isotones, are listed below. The experimental $B(E1)/B(E2)$ ratios in these nuclei were deduced as mentioned in detail in Sec. 4.2 (see also Fig. 4.8). The experimental $B(E2;\ 0^+\ \rightarrow\ 2^+)$ i.e., $B(E2)\uparrow$ values in the below mentioned nuclei were taken from [22]: $B(E2;\ 0^+ \rightarrow^+)$ for $^{144}Ba = 1.012 \pm 0.055$, $^{146}Ce = 0.97 \pm 0.11$ and $^{148}Nd = 1.338 \pm 0.030$ in $e^2 b^2$ units.

Nucleus	Spin [$\hbar$]	$B(E1)/B(E1)_w^{a)}$
^{144}Ba	7	$0.32(5) \times 10^{-3}$
		$0.33(6) \times 10^{-3}$ [b)]
	8	$1.90(30) \times 10^{-3}$
		$2.3(7) \times 10^{-3}$ [b)]
	9	$1.08(13) \times 10^{-3}$
		$0.93(12) \times 10^{-3}$ [b)]
	10	$1.61(23) \times 10^{-3}$
		$2.8(8) \times 10^{-3}$ [b)]
	11	$1.40(25) \times 10^{-3}$
		$1.5(4) \times 10^{-3}$ [b)]
^{146}Ce	7	$0.86(28) \times 10^{-3}$ [c)]
	8	$3.07(42) \times 10^{-3}$ [c)]
	9	$2.60(54) \times 10^{-3}$ [c)]
	10	$2.5(1.1) \times 10^{-3}$ [c)]
	11	$8.3(3.8) \times 10^{-3}$ [c)]
^{148}Nd	7	$2.86(48) \times 10^{-3}$
	8	$3.41(49) \times 10^{-3}$
	9	$3.24(50) \times 10^{-3}$
	10	$2.49(75) \times 10^{-3}$
	11	$4.84(1.28) \times 10^{-3}$

[a)] $B(E1)_w = 0.06446\ A^{2/3}\ e^2\ fm^2$.
[b)] See [28], [c)] see [19].

$N = 88$ isotones. As for the Ra and Th isotopes, the enhancement of the $E1$ transitions is found in these nuclei.

4.4. Electric Dipole Moments D_o

The reduced $B(E1)$ and the $B(E2)$ rates can be related to the intrinsic dipole D_o and the quadrupole Q_o moments, respectively, by the following rotational model relations [29–31]:

$$B(E1; \ I_i \rightarrow I_f) = \frac{3}{4\pi} D_0^2 \langle I_i K_i 10 | I_f K_f \rangle^2 \qquad (4.11)$$

In odd-A nuclei for $K = 1/2$ bands, a signature dependent term which has the spherical component (D_1) of the intrinsic electric dipole moment, is to be added on the right-hand side of the above expression of $B(E1)$ [32]. As discussed in [33], D_1/D_o ratio in the actinide region is small and of the order of the experimental errors in the $B(E1)/B(E2)$ ratios, the contribution of the spherical component (D_1) to $B(E1)$ has hence been ignored and the above relation used in the calculation of $B(E1)$ value,
and

$$B(E2; \ I_i \rightarrow I_f) = \frac{5}{16\pi} Q_0^2 \langle I_i K_i 20 | I_f K_f \rangle^2 \qquad (4.12)$$

According to [30], upon evaluating the Clebsch–Gordan coefficients, the expression for the ratio of the dipole D_o and the quadrupole Q_o moment can be written as

$$\left(\frac{D_0}{Q_0} \right)^2 = \frac{5}{8} \frac{B(E1)}{B(E2)} \frac{(I + K - 1)(I - K - 1)}{(2I - 1)(I - 1)} \qquad (4.13)$$

Here, the intrinsic quadrupole moment Q_o is assumed constant for the entire rotational band (i.e., not depending on spin values). Although, strictly this may not be true. Since many nuclei of interest are not good rotors, the above formula (Eq. (4.13)) which is based on the rotational model, is questionable.* But, the above is a

---

*For further reading on the validity of rotational model formula for the estimation of transition strengths in near spherical or weakly quadrupole deformed nuclei, the reader can refer to [34, 35].

consistent method for extracting the dipole moment $|D_o|$ from the $B(E1)/B(E2)$ ratio and Q_o, for decay from levels with negative and positive parities in nuclei. The data for $|D_o|$ obtained in the same way can then be compared for different nuclei of interest. Experimental value of Q_o, if available, is used in the calculation of $|D_o|$ from the above relation. If such a value is not available, then, in even-even nuclei, one may use the systematic relationship between the energy of the first 2^+ excited state to its gamma-ray lifetime, Z and A, first derived by Grodzins [36]. A number of phenomenological formulae that connect the above quantities are described in [10]. The gamma-ray lifetime is converted to $B(E2)\uparrow$ and then to Q_o through the formulae given in [10]. In [31], the following expression is used which connects Q_o, $E(2^+)$, Z and A:

$$Q_0 = 35\frac{Z}{\sqrt{E_\gamma(2_1^+ \to 0_1^+)A}} \qquad (4.14)$$

where E_γ is in MeV and Q_o in e·fm^2. The values of $B(E2)\uparrow$ (in e^{2}b^2) and Q_o (in b) for even-even nuclei are given in [10] and $B(E2)\uparrow$ (in e^{2}b^2) in [22].

The value of Q_o is calculated from $B(E2)\uparrow$ using the following relation [10]

$$Q_0 = \left[\frac{16\pi}{5}\frac{B(E2)\uparrow}{e^2}\right]^{1/2} \qquad (4.15)$$

For an odd-even and odd-odd nucleus, the Q_o value is obtained by taking an arithmetic average of Q_o values of the adjacent even-even nuclei. Adopted values of the intrinsic electric quadrupole moments in the even-even and the odd-N Ra and Th isotopes are given in Table 4.3. A plot of Q_o versus mass number for the Ra and Th nuclei is shown in Fig. 4.10.

The experimentally deduced values of the intrinsic dipole moment D_o obtained from D_o/Q_o ratio and Q_o value are listed in Tables 4.4 and 4.5 for the Ra and Th isotopes, respectively. The D_o value as a function of neutron number N are plotted in Fig. 4.11 for the Ra isotopes. The D_o values show an increasing trend from

Table 4.3. Adopted values of reduced electric quadrupole transition probability $B(E2; 0^+ \rightarrow 2^+)$, i.e., $B(E2)\uparrow$ and the intrinsic quadrupole moment Q_o
in Ra and Th isotopes.

Nucleus	$B(E2)\uparrow^*$ (e^2b^2)	$Q_o^\dagger$ (b)
^{218}Ra	1.00 ± 0.10	3.17 ± 0.16
^{219}Ra		$4.04^{a)}$
^{220}Ra		$4.92^{b)}$
^{221}Ra		$5.83^{a)}$
^{222}Ra	4.51 ± 0.36	6.73 ± 0.27
^{223}Ra		$6.53^{a)}$
^{224}Ra	3.990 ± 0.052	6.33 ± 0.04
^{225}Ra		$6.47 \pm 0.15^{c)}$
^{226}Ra	5.16 ± 0.13	7.20 ± 0.09
^{227}Ra		$7.47^{a)}$
^{228}Ra	5.98 ± 0.20	7.75 ± 0.13
^{220}Th		$3.42^{b)}$
^{221}Th		$4.44^{a)}$
^{222}Th	2.98 ± 0.25	5.47 ± 0.23
^{223}Th		$5.89^{a)}$
^{224}Th	3.96 ± 0.27	6.31 ± 0.21
^{225}Th		$7.29^{a)}$
^{226}Th	6.82 ± 0.35	8.28 ± 0.21
^{227}Th		$8.35^{a)}$
^{228}Th	7.05 ± 0.12	8.42 ± 0.07
^{230}Th	$8.04 \pm 0.10^{d)}$	$8.99 \pm 0.06^{d)}$

*These values are from [22] except as noted.
†Calc. from $B(E2)\uparrow$ (column 2) in even-even nuclei using the relation Eq. (4.15)
from [10] except as noted.
$^{a)}$ Taking arithmetic average of adopted Q_o of adjacent even-even nuclei.
$^{b)}$ Calculated using the relation Eq. (4.14) given in [31]; E_γ (^{220}Ra; $2^+ \rightarrow$
0^+) = 0.17847 MeV and E_γ (^{220}Th; $2^+ \rightarrow 0^+$) = 0.38650 MeV [3].
$^{c)}$ From [37].
$^{d)}$ From [38].

$N = 130$ to $N \sim 132$–133 where it saturates and then it decreases
with further increase in neutron number. A striking feature of this
plot is the unusually low value of D_o for $N = 136$ (^{224}Ra) nucleus.
A theoretical explanation of this feature is given later in this section.
It needs to be mentioned here that in ($N = 135$) ^{223}Ra nucleus, in

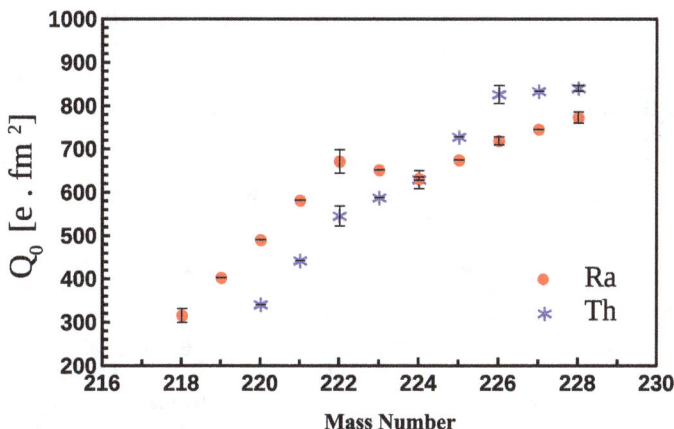

Fig. 4.10. Plot of intrinsic quadrupole moment, Q_o versus mass number for Ra (•) and Th (∗) isotopes. See Table 4.3 for details on experimental data.

the $K = 1/2$ band for $I < 8$, $D_o = 0.078 \pm 0.012$ e.fm. This is also a relatively low value of the intrinsic dipole moment. For discussion on the K-dependence of D_o in this nucleus, see [39].

In Fig. 4.12 the intrinsic dipole moment D_o is plotted as a function of neutron number N for the Th isotopes. This plot also shows the increasing trend in D_o from $N = 130$ to $N = 134$ where it saturates and then decreases with the increase in neutron number. No abnormal behavior in D_o is observed for the Th isotopes.

Let us now discuss the electric dipole moments D_o in some of the neutron rich even-even Ba isotopes. We first consider the intrinsic quadrupole moments Q_o in these nuclei. Table 4.6 gives these values. A plot of Q_o versus mass number for the Ba isotopes is shown in Fig. 4.13. It shows a monotonous increase in the Q_o values with the increase in mass number.

Figure 4.14 gives a plot of the experimental values of electric dipole moment D_o from [18] as a function of mass number, for the isotopes of Ba from $N = 84$ to 92. In the above work, the $B(E1)/B(E2)$ ratios were first obtained from the measurement of the γ-branching ratios and the γ-ray energies in these nuclei. An average of $B(E1)/B(E2)$ ratios was then taken for each nucleus for spins $I \geq 6$. Using the averaged value of $B(E1)/B(E2)$ for

Table 4.4. Experimental values of intrinsic dipole moment $|D_o|$ (column 5) for Ra ($Z = 88$) isotopes. These are derived from the intrinsic quadrupole moment Q_o (column 2) and the weighted average experimental value of $|D_o/Q_o|$ (column 4). The D_o values so obtained are also compared with earlier works. The K-value for the bands and the spin I ranges are mentioned in column 3.

Nucleus	$Q_o^{a)}$ (e. fm^2)	K value, I range	wt. av. $|D_o/Q_o|^{d)}$ ($10^{-4} \cdot$ fm^{-1})	$|D_o|$ (e · fm)
^{218}Ra$_{130}$	317 ± 16	$K = 0, I = 10 - 15$	6.48 ± 0.31	0.21 ± 0.01
		$I = 6$		$0.23 \pm 0.05^{g)}$
		$I = 7 - 11$		$0.339 \pm 0.017^{g)}$
^{219}Ra$_{131}$	$404^{b)}$	$K = 1/2, I = 21/2 - 41/2$	7.05 ± 0.10	0.28 ± 0.03
		$I = 19/2 - 51/2$		$0.32 \pm 0.03^{g)}$
^{220}Ra$_{132}$	$492^{b)}$	$K = 0, I = 7 - 28$	7.68 ± 0.19	0.38 ± 0.04
		$I = 7 - 17$		$0.27 \pm 0.07^{g)}$
^{221}Ra$_{133}$	$583^{b)}$	$K = 5/2, I = 17/2 - 25/2$		$0.36 \pm 0.10^{g)}$
^{222}Ra$_{134}$	$674 \pm 28^{c)}$	$K = 0, I = 7 - 15$	$4.02 \pm 0.11^{e)}$	0.27 ± 0.01
^{223}Ra$_{135}$		$K = 1/2, I < 8$		$0.078 \pm 0.012^{h)}$
		$K = 3/2, I < 8$		$0.155 \pm 0.010^{i)}$
		$I = 3/2 - 11/2$		$0.124 \pm 0.010^{h)}$
		$K = 5/2, I < 8$		$0.043 \pm 0.012^{h)}$
^{224}Ra$_{136}$	633 ± 4	$K = 0, I = 5 - 7$	$0.47 \pm 0.02^{e)}$	0.030 ± 0.001
		$K = 0, I = 3 - 5$		$0.028 \pm 0.004^{g)}$
		$K = 0, I = 3, 5$		$0.0296 \pm 0.0009^{j)}$
^{225}Ra$_{137}$		$K = 1/2, I = 3/2$		$0.161 \pm 0.008^{g)}$
^{226}Ra$_{138}$	720 ± 9	$K = 0, I = 9 - 18$	$2.39 \pm 0.08^{e)}$	0.172 ± 0.006
		$I = 7 - 11$	$2.14 \pm 0.07^{f)}$	$0.154 \pm 0.005^{k)}$
^{227}Ra$_{139}$		$K = 3/2$		$0.098 \pm 0.011^{h)}$
^{228}Ra$_{140}$	775 ± 13	$K = 0, I = 11 - 17$	$1.03 \pm 0.13^{e)}$	0.08 ± 0.01

(a) See Table 4.3, except as noted.

(b) 10% error in Q_o assumed.

(c) Adopted from [38].

(d) Calculated from $B(E1)/B(E2)$ ratios (see Figs. 4.1 and 4.2) using Eq. (4.13), except as noted.

(e) Adopted from [7].

(f) Adopted from [13].

(g) Quoted in [31].

(h) Adopted from [32].

(i) Average D_o determined from experimental $B(E1)$ rates obtained through lifetime measurements for 50.1 and 79.7 keV levels [39].

(j) From Coulomb excitation experiment [40]; see also [41].

(k) Using wt. av. $|D_o/Q_o|$ from [13] mentioned in column 4 and Q_o given in column 2.

Table 4.5. Experimental values of intrinsic dipole moment $|D_o|$ (column 5) for Th ($Z = 90$) isotopes. These are derived from the intrinsic quadrupole moment Q_o (column 2) and the weighted average experimental value of $|D_o/Q_o|$ (column 4). The D_o values so obtained are also compared with earlier works. The K-value for the bands and the spin I ranges are mentioned in column 3.

Nucleus	$Q_o^{a)}$ (e. fm^2)	K value, I range	wt. av. $\|D_o/Q_o\|^{d)}$ ($10^{-4} \cdot$ fm^{-1})	$\|D_o\|$ (e · fm)
^{220}Th$_{130}$	342$^{b)}$	$K = 0, I = 6 - 12$	8.59 ± 0.07	0.29 ± 0.03
		$K = 0, I = 6 - 11$		$0.25 \pm 0.03^{f)}$
^{221}Th$_{131}$	444$^{b)}$	$K = 1/2, I = 15/2 - 39/2$	8.03 ± 0.09	0.36 ± 0.04
		$I = 23/2 - 35/2$		$0.31 \pm 0.02^{g)}$
		$I = 17/2 - 37/2$		$0.33 \pm 0.06^{h)}$
^{222}Th$_{132}$	547 ± 23	$K = 0, I = 6 - 17$	6.7 ± 0.3	0.37 ± 0.02
^{223}Th$_{133}$	589$^{b)}$	$K = 5/2, I = 15/2 - 45/2$		$0.41 \pm 0.03^{i)}$
		$K = 5/2, I = 11/2 - 33/2$		$0.44 \pm 0.09^{h)}$
^{224}Th$_{134}$	631 ± 21	$K = 0, I = 6 - 17$	$8.4 \pm 0.3^{e)}$	0.53 ± 0.03
^{225}Th$_{135}$	722$^{b)\ c)}$	$K = 3/2, I = 15/2 - 31/2$	$5.1 \pm 0.6^{c)}$	$0.37 \pm 0.05^{c)}$
^{226}Th$_{136}$	828 ± 21	$K = 0, I = 8 - 19$	$3.7 \pm 0.1^{e)}$	0.30 ± 0.01
^{227}Th$_{137}$	835$^{b)}$	$K = 1/2$		$0.21 \pm 0.03^{j)}$
^{228}Th$_{138}$	842 ± 7	$K = 0, I = 9 - 13$	$1.44 \pm 0.03^{e)}$	0.121 ± 0.003
^{230}Th$_{140}$	899 ± 6	$K = 0, I = 3 - 9$	$0.59 \pm 0.02^{e)}$	0.053 ± 0.002

$^{(a)}$See Table 4.3, except as noted.

$^{(b)}$10% error in Q_o assumed.

$^{(c)}$From [1].

$^{(d)}$Calculated from $B(E1)/B(E2)$ ratios (see Figs. 4.4 and 4.5) using Eq. (4.13), except as noted.

$^{(e)}$From [13].

$^{(f)}$See [31].

$^{(g)}$Priv. comm. from Dr. Sujit Tandel (March 2017); see also [42].

$^{(h)}$From [30].

$^{(i)}$Re-calculated value from $B(E1)/B(E2)$ ratios obtained from the experimental γ-branching ratios [17] and Q_o given in column 2 in this table assuming 10% error in Q_o. The weighted average $|D_o/Q_o|$ value obtained from $B(E1)/B(E2)$ is $|D_o/Q_o| = 7.22 \pm 0.33$ ($10^{-4} \cdot$ fm^{-1}) for $s = -i$ band and $|D_o/Q_o| = 6.63 \pm 0.33$ ($10^{-4} \cdot$ fm^{-1}) for $s = +i$ band. The $|D_o|$ value mentioned in column 5 is weighted mean of the $|D_o|$ values obtained for the two bands in this nucleus.

$^{(j)}$From [33].

each nucleus, the D_o value was then obtained using the approximate relation

$$D_0 = \sqrt{\frac{5}{16} \frac{B(E1)}{B(E2)}} Q_0 \qquad (4.16)$$

Pear-Shaped Nuclei

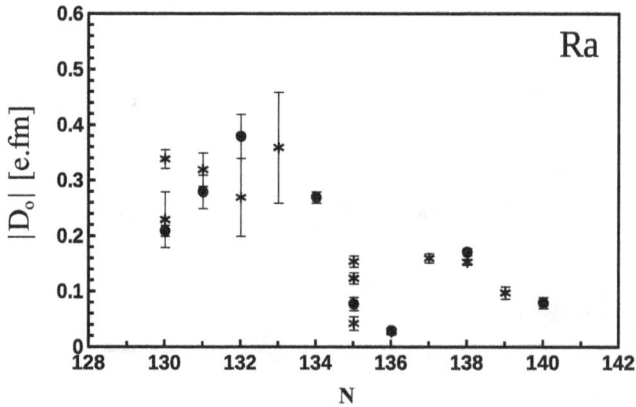

Fig. 4.11. Electric dipole moment $|D_o|$ as a function of neutron number N for the Ra isotopes. Experimental data are taken from sources mentioned in Table 4.4. For the data points marked with (•), the values are calculated as mentioned in the title of the table except $D_o = 0.078 \pm 0.012$ (e.fm) for the $K = 1/2$ band in ^{223}Ra (see footnote "h" in the table). The other data points marked with (∗) are from earlier works (see the corresponding footnotes in the table).

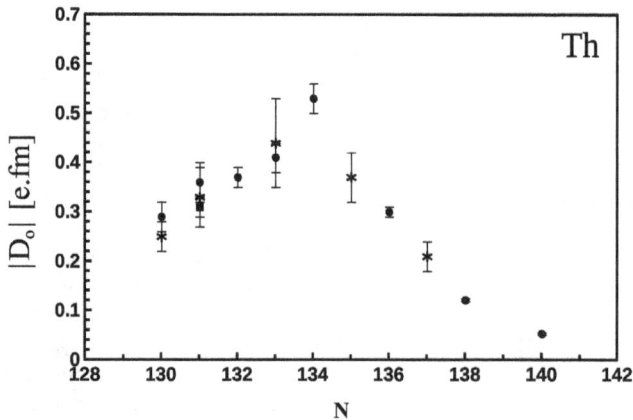

Fig. 4.12. Electric dipole moment $|D_o|$ as a function of neutron number N for the Th isotopes. Experimental data are taken from sources mentioned in Table 4.5. For the data points marked with (•), the values are calculated as mentioned in the title of the table and the data points with (∗) markings are from earlier works. For (■) see footnote "g" in the Table.

Table 4.6. Adopted values of reduced electric quadrupole transition probability $B(E2; 0^+ \rightarrow 2^+)$, i.e., $B(E2)\uparrow$, quadrupole deformation parameter β_2 and the intrinsic quadrupole moment Q_o in the Ba isotopes from $N = 84 - 92$.

Nucleus	$B(E2)\uparrow^*$ (e^2b^2)	β_2^*	$Q_o^\dagger$ (b)
^{140}Ba	0.484 ± 0.038	0.1340 ± 52	2.21 ± 0.09
^{142}Ba	0.676 ± 0.035	0.1569 ± 41	2.61 ± 0.07
^{144}Ba	1.012 ± 0.055	0.1902 ± 52	3.19 ± 0.09
^{146}Ba	1.350 ± 0.068	0.2177 ± 55	3.68 ± 0.09
^{148}Ba			4.28 ‡

*These values are from [22]; $\beta_2 = (4\pi/3ZR_o^2) \, [B(E2)\uparrow/e^2]^{1/2}$ and $R_o^2 = (1.2 \times 10^{-13} \, A^{1/3} \, \text{cm})^2$.
†Calculated from $B(E2)\uparrow$ (column 2) using the relation Eq. (4.15) from [10] unless noted otherwise.
‡Calculated using the relation Eq. (4.15) in [31].

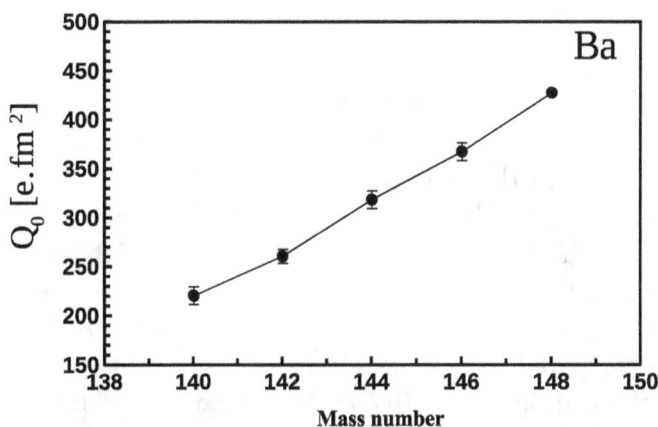

Fig. 4.13. Plot of intrinsic quadrupole moment, Q_o versus mass number for the even-even Ba isotopes from $A = 140$ to 148. See Table 4.6 for the source from which the experimental data are taken.

and the quadrupole moment Q_o values from [38]. As Q_o for ^{142}Ba and ^{148}Ba nuclei are not available in [38], $Q_o = 2.1(3)$ e.b for ^{142}Ba and $Q_o = 4.2(8)$ e.b for ^{148}Ba were adopted by the authors. Comparing the obtained D_o values in the ^{140}Ba, ^{142}Ba, ^{144}Ba, ^{146}Ba and ^{148}Ba

Pear-Shaped Nuclei

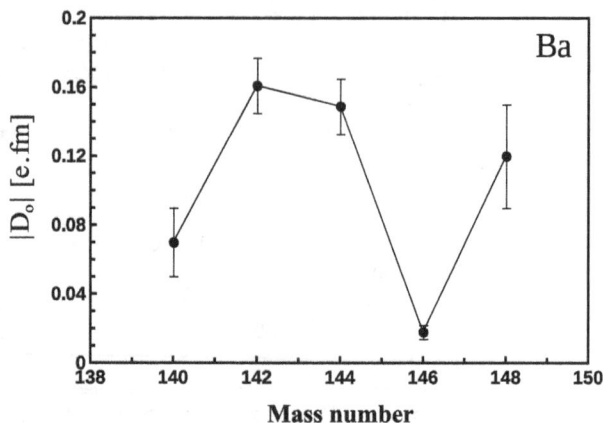

Fig. 4.14. Intrinsic electric dipole moments $|D_0|$ as determined experimentally in [18] (the numerical values of D_0 are as read from Fig. 11 of this work and communicated by [44], in the even-even ^{140,142,146,148}Ba isotopes. For D_0 values obtained in other measurements in the even-even Ba isotopes, see [28, 45–47] and also see [31, 32].

nuclei, it is seen from this figure (Fig. 4.14) that there is a sudden drop in the value of D_0 in ^{146}Ba ($N = 90$). This peculiarity was first found between ^{144}Ba and ^{146}Ba nuclei in [28]. The D_0 value returns quickly in the neighboring even-even nucleus ^{148}Ba near to the level as seen in ^{144}Ba. If one looks at the experimental values of the $B(E1)/B(E2)$ ratios obtained in [18] in ^{144}Ba, these are of the order of 10^{-6} fm^{-2}. But in ^{146}Ba, these are almost two orders of magnitude smaller than in ^{144}Ba. The increase in the $B(E2)$ value in ^{146}Ba as compared to that in ^{144}Ba (see Table 4.6) cannot account for this sharp drop in the $B(E1)/B(E2)$ ratio in ^{146}Ba. This drop in the $B(E1)/B(E2)$ ratio in ^{146}Ba is most likely due to a sudden decrease in the $E1$ transition strength. The $E1$ transitions in ^{146}Ba are not observed above spin 9 in the negative parity band based on the 821.3 keV 3^- state (see Fig. 4.15). A theoretical explanation of this problem will be discussed later in this section.

We now talk about the electric dipole moments D_0 in the $N = 88$ isotones. In Table 4.7 the intrinsic quadrupole moments Q_0 in these nuclei are given. In Table 4.8, in the last column, are given the experimentally deduced intrinsic dipole moments D_0 for

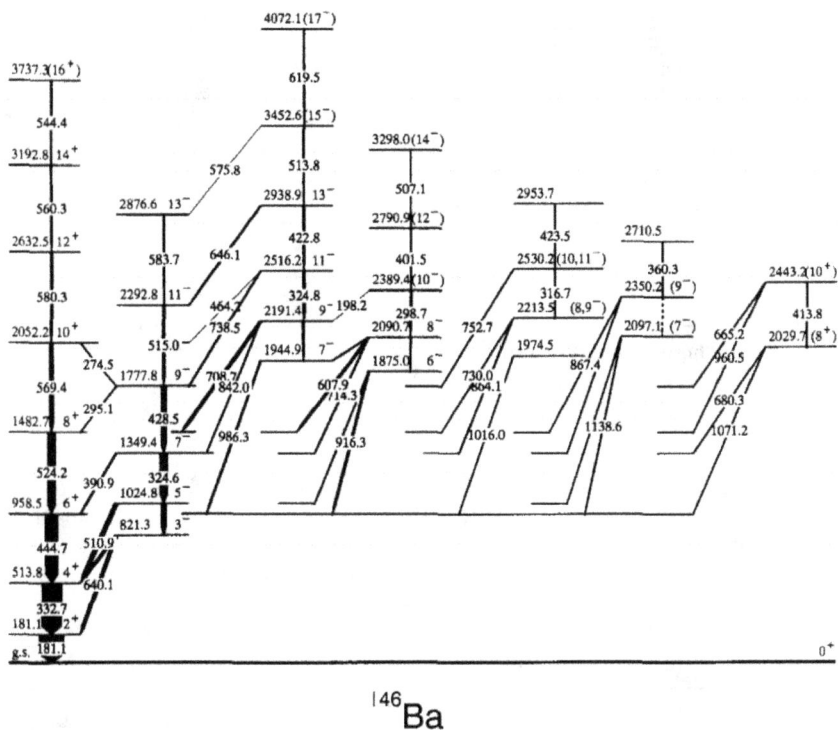

Fig. 4.15. Level scheme of ^{146}Ba as obtained in [18]. Figure is reproduced with permission from [18].

the ^{144}Ba, ^{146}Ce, ^{148}Nd and ^{150}Sm isotones. It is found that there is an increase of D_0 from ^{144}Ba ($Z = 56$) to ^{146}Ce ($Z = 58$) to ^{148}Nd ($Z = 60$) peaking at $Z = 60$. For a discussion on the detailed proton dependence of the dipole moment D_0 in nuclei from $Z = 56$ to 65, see [43]. Their Fig. 3 which is a plot of the electric dipole moment (the notation Q_1 has been used there instead of D_0) versus proton number, shows the maximum value of the dipole moment at also around $Z = 60$.

There has been a lot of interest to interpret the experimentally obtained values of intrinsic electric dipole moments D_0 and their variation with nucleon number in nuclei exhibiting octupole correlations. The electric dipole moments were first theoretically calculated considering the liquid drop model alone in [48–51]. According to

Table 4.7. Adopted values of reduced electric quadrupole transition probability $B(E2; 0^+ \rightarrow 2^+)$, i.e., $B(E2)\uparrow$ and the intrinsic quadrupole moment Q_o in $N = 88$ nuclei.

Nucleus	$B(E2)\uparrow^*$ (e^2b^2)	$Q_o^\dagger$ (b)
^{144}Ba	1.012 ± 0.055	3.19 ± 0.09
^{146}Ce	0.97 ± 0.11	3.12 ± 0.18
^{148}Nd	1.338 ± 0.030	3.67 ± 0.04
^{150}Sm	1.347 ± 0.026	3.68 ± 0.03

*These values are from [22].
†Calc. from $B(E2)\uparrow$ (column 2) in even-even nuclei using the relation Eq. (4.15) from [10].

Table 4.8. Experimental values of intrinsic electric dipole moment $|D_o|$ for $N = 88$ nuclei. These are derived from the intrinsic quadrupole moment Q_o (column 2) and the weighted average experimental value of $|D_o/Q_o|$ (column 4). $K = 0$ for the bands. The spin I range is mentioned in column 3.

Nucleus	Q_o^* (e. fm^2)	Spin range	wt. av. $|D_o/Q_o|$ $(10^{-4}.$ fm$^{-1})$	$|D_o|$ (e. fm)
^{144}Ba	319 ± 9	$I = 8 - 16$	$4.56 \pm 0.16^{a)}$	0.145 ± 0.006
		$I = 8 - 11$		$0.14 \pm 0.03^{b)}$
^{146}Ce	312 ± 18	$I = 8 - 10$	$6.34 \pm 0.48^{c)}$	0.20 ± 0.02
^{148}Nd	367 ± 4	$I = 7 - 10$	$6.09 \pm 0.26^{d)}$	0.22 ± 0.01
		$I = 5 - 8$		$0.24 \pm 0.03^{b)}$
^{150}Sm	368 ± 3	$I = 16 - 22$	$4.60^{e)}$	0.17 ± 0.02
		$I = 7 - 15$		$0.19 \pm 0.03^{b)}$

*See Table 4.7.
$^{(a)}$Obtained from $B(E1)/B(E2)$ values derived from the experimental γ-branching ratios [18].
$^{(b)}$Quoted in [32].
$^{(c)}$Obtained from $B(E1)/B(E2)$ values from [19].
$^{(d)}$Obtained from $B(E1)/B(E2)$ values from [20].
$^{(e)}$Obtained from $B(E1)/B(E2) = 0.7 \ 10^{-6}$ fm^{-2} from [14].

Strutinski, for a liquid drop with octupole deformation, the Coulomb potential will push the protons towards regions of maximum curvature. This effect leads to density polarization of protons resulting in the displacement between the proton and neutron centers of mass.

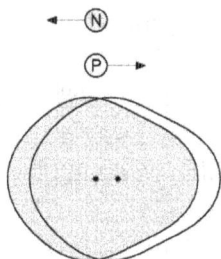

Fig. 4.16. Schematic depiction of the displacement of the proton and neutron centers of mass due to Coulomb volume polarization in a liquid drop with octupole deformation ($\beta_3 < 0$ or $\varepsilon_3 > 0$).

This displacement is schematically shown in Fig. 4.16. Therefore, the axially symmetric reflection asymmetric nuclei have intrinsic (polarized) electric dipole moment. Since these early calculations, in the last 30 over years, a large number of theoretical efforts have been made to understand the occurrence and behavior of electric dipole moments in such nuclei [31, 45, 52–65].

The liquid drop formalism could not explain the experimental findings that the $B(E1)/B(E2)$ ratios are about two orders of magnitude smaller in ^{224}Ra and ^{146}Ba as compared to their $(N-2)$ isotopes, namely ^{222}Ra and ^{144}Ba (see [7] and [18] respectively). In all the four above-mentioned nuclei, the octupole deformation is well-developed, $\beta_3 \approx 0.1$ [31]. A shell effect or shell correction term was therefore added to the liquid drop term by Leander [52] which takes into account the effect of shell structure on the dipole moment, to explain the above mentioned behavior. Detailed intrinsic dipole moment D_0 calculations based on this macroscopic (liquid drop)–microscopic (shell correction) approach were done in [31]. The intrinsic dipole moment D_0 within this approach [31, 54] can be written as

$$D_0^{\text{total}} \ (\text{or } D_0) = D_0^{\text{macro}} + D_0^{\text{shell}} \tag{4.17}$$

where D_0^{macro} is the macroscopic contribution and D_0^{shell} is the shell correction contribution. The magnitude of the D_0^{macro} contribution depends upon the signs and magnitudes of the charge redistribution and neutron skin terms. Similarly, in D_0^{shell} one should consider the signs and magnitudes of the proton and the neutron terms. Overall, the signs and magnitudes of the D_0^{macro} and D_0^{shell} terms determine the total intrinsic dipole moment D_0^{total}. Numerical values of these two contributions and the total D_0^{total} (or D_0) are tabulated for a large number of nuclei exhibiting octupole correlations in the actinide and the lanthanide regions in [31].

In Fig. 4.17, the calculated [31] D_0^{macro}, D_0^{shell} and the total D_0^{total} values are plotted as a function of neutron number N, for the even-even isotopes of Ra (upper panel) and Th (lower panel) in the range $N = 130$ to 138 (i.e., ^{218}Ra to ^{226}Ra and ^{220}Th to ^{228}Th nuclei respectively). In ^{224}Ra ($N = 136$), the small experimental value of D_0 (see Table 4.4 and Fig. 4.11) is explained as a result of the cancellation of the D_0^{macro} and the D_0^{shell} contributions. For the $N = 130$, 132 and 134 isotopes of Ra, positive and finite values of D_0 are predicted by theory. However, in $N = 138$ (^{226}Ra) a finite negative D_0 is predicted. In the odd-N ^{223}Ra ($N = 135$) nucleus, in $K = 1/2$ band, the low experimental value of D_0 can also be explained by substantial cancellation between the macro and the micro contributions (see Table 6 in [31] for theoretical values of dipole moments D_0^{macro} and D_0^{shell} in ^{223}Ra).

For the Th isotopes, the theory predicts (see lower panel in Fig. 4.17) the general trend found experimentally (see Fig. 4.12) in the D_0 values as a function of neutron number. In [64] detailed theoretical calculations have been performed for the macroscopic contribution, D_0^{macro} to the total electric dipole moment D_0^{total}. In this work, the electric dipole moment has been named as *polarized* electric dipole moment. A detailed expression for the macroscopic contribution has been derived for the case of geometrically similar proton and neutron surfaces within the framework of the liquid drop model in well-deformed reflection asymmetric nuclei. It has been shown that the contributions of the second-order terms to the expression are important for well-deformed nuclei. In Fig. 4.18 [64], the results of the calculations are plotted for the Th isotopes. In the work, the deformation parameters and the D_0^{shell} part have been adopted from [31]. The experimental D_0 values (see also Fig. 4.12) as a function of mass number are compared with various theoretical

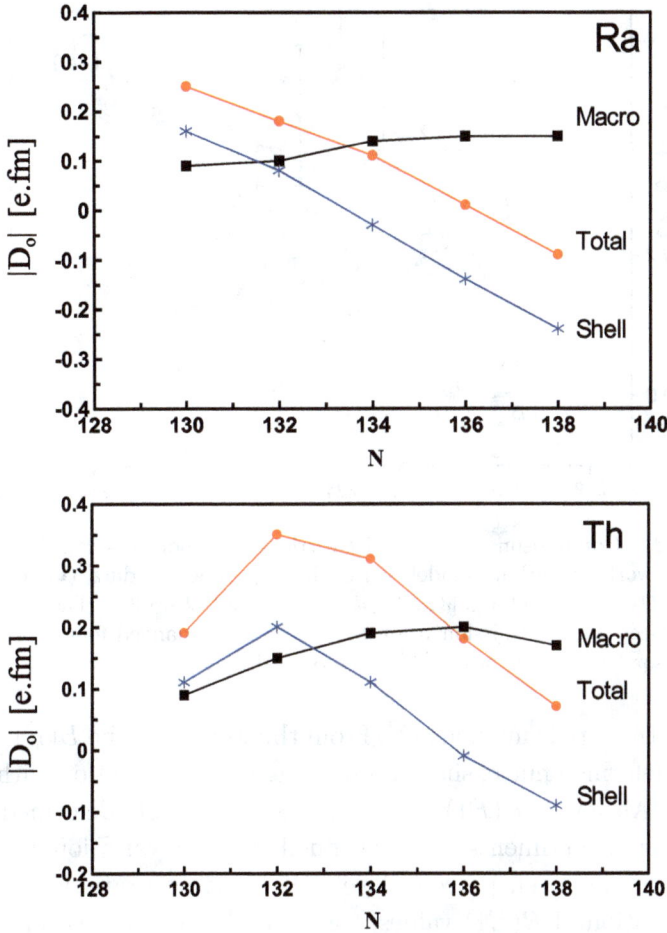

Fig. 4.17. Values of macroscopic contribution, D_o^{macro} (■), shell-correction contribution, D_o^{shell} (∗) and the total intrinsic electric dipole moment, Do^{total} (●)[= $D_o^{macro} + D_o^{shell}$] calculated in [31] using the macroscopic-microscopic model are shown as a function of neutron number N, for the even-even isotopes of Ra (top panel) and Th (bottom panel).

results. It is seen from the figure (Fig. 4.18) that the D_o^{macro} values calculated from the work [64] plus the D_o^{shell} values from the work in [31] is in good agreement with the experimental data.

We note from the expression, Eq. (4.11) that the reduced electric dipole transition probability $B(E1)$ is proportional to the squared

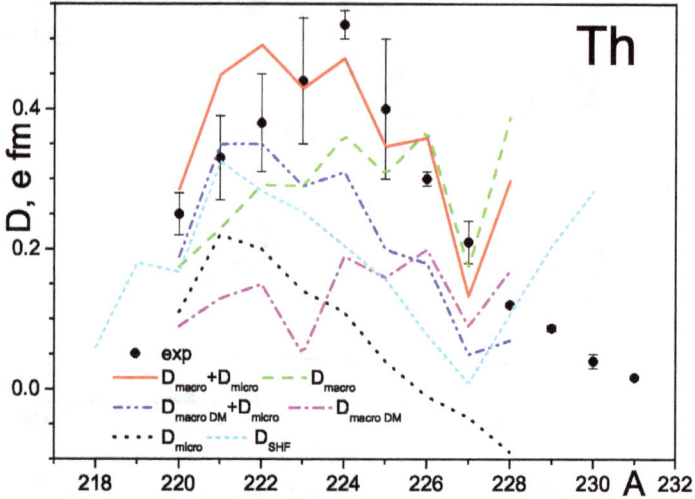

Fig. 4.18. Comparison of polarized electric dipole moments calculated within the framework of various models with the experimental data (•) for the Th isotopes. D_{macro} [64], D_{micro} [31], D_{macroDM} [31] and D_{SHF} [61]. The figure above is reproduced with permission from [64] but is with changed font type. Thanks to Professor Denisov for providing this modified figure.

value of the dipole moment D_o^2. From the work [64], the $B(E1)$ values in well-deformed nuclei should show a significant variation with deformation. Although $B(E1)$ values in many octupole deformed nuclei have been experimentally determined, such a variation in $B(E1)$ values has not been perceived yet. We need accurate experimental determination of $B(E1)$ values, e.g., from lifetime measurements, to verify the theoretical predictions.

Let us now consider the dipole moments in Ba nuclei. The theoretical values [31] of D_o^{macro}, D_o^{shell} and the total D_o^{total} for the even-even isotopes of Ba in the neutron number range $N = 84$ to 92 are plotted in Fig. 4.19. The small experimental value of $D_o = 0.018 \pm 0.004$ e.fm. found in ^{146}Ba (see Fig. 4.14) can be interpreted in this macroscopic-microscopic model through the cancellation of the proton and neutron terms in the shell correction contribution D_o^{shell}. The macroscopic contribution D_o^{macro} is already very small $D_o^{\text{macro}} \lesssim 0.03$ e.fm. in the lanthanide region of nuclei. This theory, however, cannot fully account for the level of enhancement

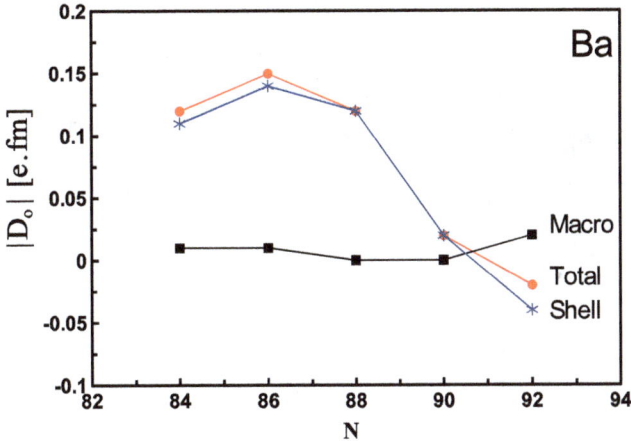

Fig. 4.19. Values of macroscopic contribution, D_o^{macro} (■), shell-correction contribution, D_o^{shell} (∗) and the total intrinsic electric dipole moment, D_o^{total} (●)[= $D_o^{\text{macro}} + D_o^{\text{shell}}$] calculated in [31] using the macroscopic-microscopic model are shown as a function of neutron number N, for the even-even isotopes of Ba from $N = 84$ to 92.

in the experimental D_0 value in ^{148}Ba which has two neutrons more than in ^{146}Ba.

The microscopic self-consistent Hartree–Fock–Bogoliubov (HFB) calculations with the Gogny interaction by [58, 59] were also able to explain the experimentally obtained low values of the intrinsic electric dipole moments D_0 in ^{224}Ra and ^{146}Ba nuclei. Recently, more elaborate microscopic self-consistent calculations have been done [65, 66]. These calculations also indicate the observed experimental trends in the dipole moments in the Ba isotopes, e.g., D_0 has positive value for ^{144}Ba, very small for ^{146}Ba and negative value for ^{148}Ba. Further, theoretical calculations [45, 59] of dipole moments as a function of octupole moment or β_3, for the even-even ^{142,144,146,148}Ba isotopes provide very interesting results. A plot of D_0^{SP} as a function of octupole deformation parameter β_3 is shown in Fig. 4.20 [45] for these Ba isotopes. In ^{142,144,148}Ba, the dipole moment varies with octupole deformation parameter but in ^{146}Ba, it is small in a range of β_3. Since only the number of neutrons change between one isotope to the other, with the proton number remaining the

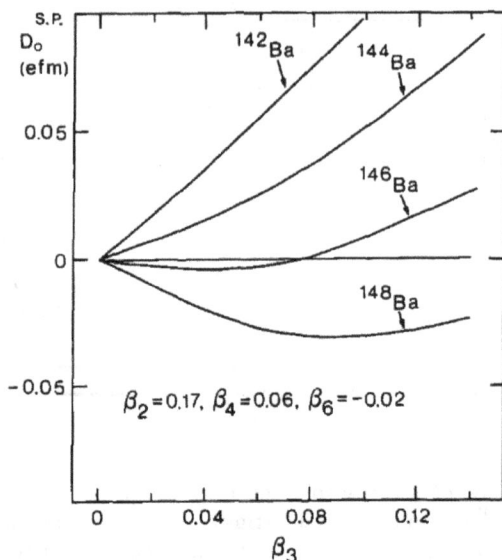

Fig. 4.20. Theoretical values of intrinsic dipole moments D_o^{SP} within the single particle model [45] as a function of octupole deformation parameter β_3 for the even-even ^{142,144,146,148}Ba isotopes for the adopted values of the other deformation parameters as mentioned in the figure frame. Figure is reproduced with permission from [55].

same ($Z = 56$), this particular behavior of dipole moment can be ascribed to the change in neutron shell contribution to total D_o^{shell} from one isotope to the other, and the occupancy of the last few neutrons in different specific single particle neutron states near the Fermi level [59, 65]. Figure 4.21 [65] is a plot of neutron single particle energies as a function of β_3 (right panel) for ^{146}Ba. The three single particle orbitals of concern near the Fermi level are with $K^\pi = 3/2^-$, $5/2^-$ and $1/2^+$ with $h_{9/2}$, $f_{7/2}$ and $i_{13/2}$ shell model parentage, respectively, having significiant occupancies. The contribution of neutrons to D_o^{shell} due to the occupation of neutrons in these orbitals results in almost cancellation as that due to protons. Since D_o^{macro} is already small for these isotopes, the total D_o^{total} becomes small. In the case of ^{144}Ba, these orbitals are empty but in ^{148}Ba with two additional neutrons than in ^{146}Ba, D_o changes sign and returns to a significient value.

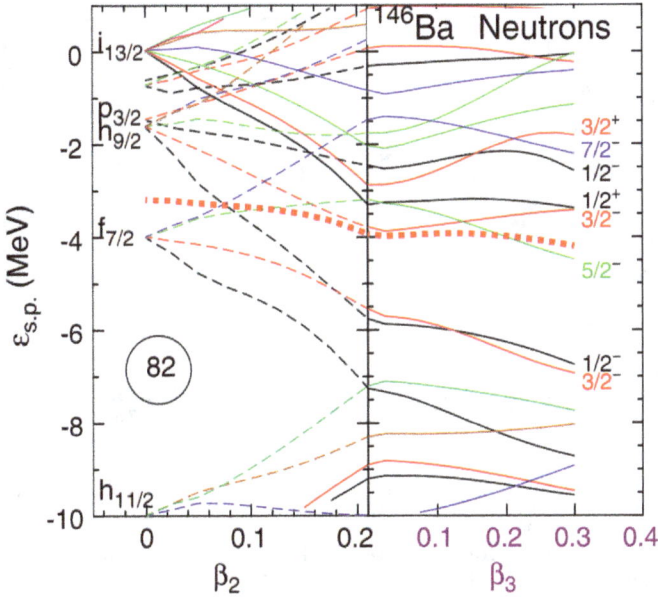

Fig. 4.21. Plot of neutron single particle energies, $\varepsilon_{s.p.}$, as a function of quadrupole deformation parameter, β_2 (left side) and as a function of octupole deformation parameter, β_3 (right side). At $\beta_2 = 0$ the spherical shell model orbitals are mentioned. For the β_3 plot, a constant quadrupole deformation value was fixed at $\beta_2 = 0.2$ (the potential energy surface minimum). The central vertical axis is at $\beta_3 = 0$. The thick line (∎∎∎), depicts on both sides, the Fermi level. On the extreme right are mentioned the K^π quantum numbers. The original figure is courtesy of Dr. B. Bucher. Figure is reproduced with permission from [65].

4.5. Excitation Energy of 3⁻ States

The excitation energy of 3^- states $E(3^-)$ in nuclei can be used as an indicator of the strength of octupole collectivity in nuclei [67–70]. Here, Fig. 4.22 shows a plot of the energy of the lowest lying 3^- states relative to the 0^+ ground state in the even-even Rn, Ra, Th and U and Pu nuclei. Even if the nature of variation of $E(3^-)$ with neutron number is similar in both the Ra-Th and the U-Pu regions of nuclei, the minima of $E(3^-)$ are much shallower for the heavier nuclei in comparison to the deep minima in Ra-Th region. The energy $E(3^-) = 317.29 \, \text{keV}$ in ^{222}Ra ($N = 134$), $290.35 \, \text{keV}$ in ^{224}Ra ($N = 136$), $321.54 \, \text{keV}$ in ^{226}Ra ($N = 138$); $305.3 \, \text{keV}$ in ^{224}Th ($N = 134$),

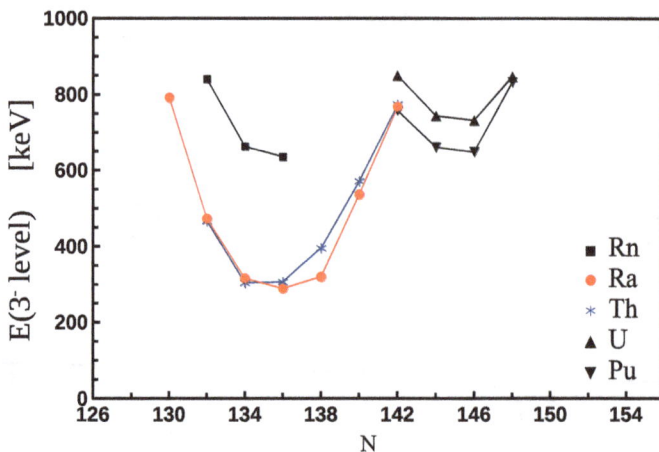

Fig. 4.22. Plot of energies of the lowest 3^- states $E(3^-)$ versus neutron number, N, in the even-even Rn, Ra and Th isotopes in the range $N = 130$ to 142 and for the U and Pu isotopes from $N = 142$ to 150. The experimental data are taken from [3]. This figure is similar to Fig. 4(a) in [70].

307.5 keV in ^{226}Th ($N = 136$) and 396.09 keV in ^{228}Th ($N = 138$) nuclei. The energy $E(3^-) = 648.86$ keV for ^{240}Pu nucleus. It may be mentioned here that stable octupole deformation has been found for ^{224}Ra and also for ^{226}Ra (see Sec. 4.6).

In Fig. 4.23, the energy of the lowest lying 3^- states $E(3^-)$ in the Ba, Ce and Nd even-even nuclei are plotted as a function of neutron number, N. These plots are not that spectacular as those for the actinides. The energies of 3^- states in Ba nuclei, ^{144}Ba ($N = 88$), $E(3^-) = 838.37$ keV and ^{146}Ba ($N = 90$), $E(3^-) = 821.10$ keV are the lowest. In these two nuclei strong octupole correlations have also been found (see Sec. 4.6).

4.6. $B(E3: \ 0^+ \to 3^-)$, Q_3 Moments and Stable Octupole Shapes

A compilation of experimental values of reduced electric octupole transition probabilities $B(E3: 0^+ \to 3^-)$ (also written as $B(E3)\uparrow$) from the ground state to the first excited 3^- state for even-even nuclei is available in literature [71]. The $E3$ single particle transition

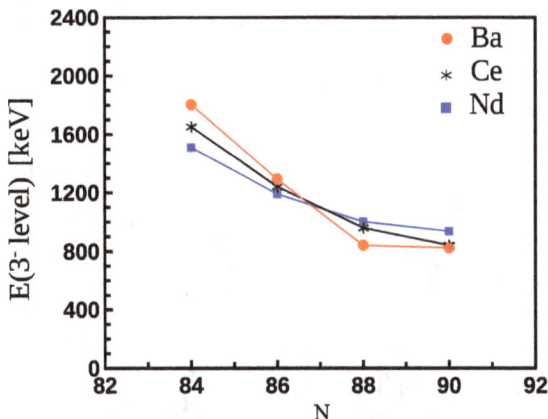

Fig. 4.23. Plot of energies of the lowest 3^- states $E(3^-)$ versus neutron number, N, in the even-even isotopes of Ba, Ce and Nd with $N = 84$ to 90. The experimental data are taken from [3].

strength $|M(E3)|^2$ in Weisskopf units (W.u.) can be obtained from the $B(E3 : 0^+ \rightarrow 3^-)$ value using the expression [71]

$$|M(E3)|^2 = 2.404 \times 10^6 B(E3)\uparrow/A^2 \quad \text{W.u.} \tag{4.18}$$

where $B(E3)\uparrow$ is in units of $e^2 b^3$ and $r_o = 1.20$ fm.

Plots of $|M(E3)|^2$ versus neutron number N (upper panel) and proton number Z (lower panel) from the above compilation are shown in Fig. 4.24. Enhanced octupole transition strengths or octupole collectivity is evident in the plot when $N \sim 34, 56, 88$ and 134. This is in agreement with lowest energies of the first 3^- states found at $N = 88, 90$ in Ba and $N = 134, 136$ in Ra, Th nuclei (see previous section, Sec. 4.5). The enhanced octupole transition strengths at these nucleon numbers agree with the theoretical predictions of the likely nuclear regions for static octupole deformation [72]. The $|M(E3)|^2$ plot versus the proton number is, however, not so clear as that for the neutron number but it does show maxima at $Z = 30$, 40, 62 and 88.

The experimental values of $B(E3 : 0^+ \rightarrow 3^-)$ are determined through a variety of experimental methods, in model independent

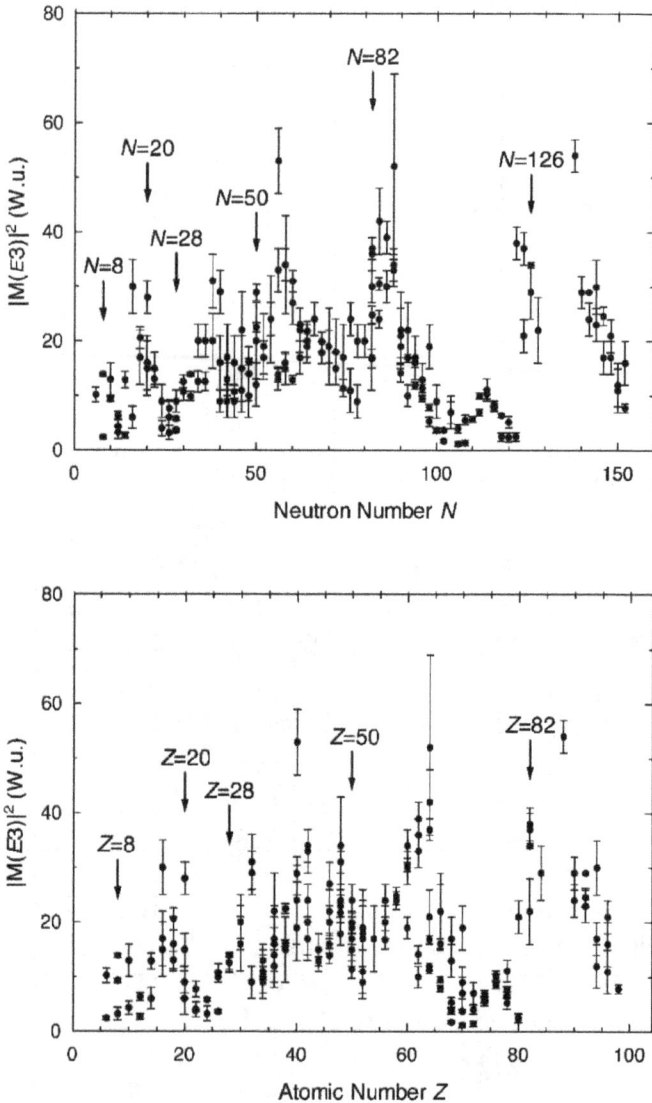

Fig. 4.24. Plots of single particle $E3$ transition strength, $|M(E3)|^2$ obtained from experimental value of $B(E3 : 0^+ \to 3^-)$ rates, as a function of neutron number N (upper panel) and atomic (proton) number Z (lower panel) in even-even nuclei. The shell model nucleon magic numbers are marked. The figures are reproduced with permission from [71].

way from Coulomb excitation of stable nuclei or lifetime measurement of nuclear states. The $B(E3 : 0^+ \rightarrow 3^-)$ values have also been obtained from inelastic electron scattering but the results are model dependent. Still another method that has been widely used is through the determination of octupole deformation parameter, β_3 from the measurement of angular distributions of inelastically scattered light particles, like, protons, neutrons, deuterons, etc. The data analysis in such experiments are model dependent. The above mentioned compilation [71] gives the listings of the values of $B(E3 : 0^+ \rightarrow 3^-)$ obtained from each of the mentioned methods with the relevant references and the adopted value for each even-even nucleus.

In [73], a theory was developed to investigate globally the lowest octupole excitations in axially symmetric even-even nuclei. In this work, $B(E3)$ values were also predicted in heavy nuclei Rn, Ra, Th, U and Pu as a function of mass number from $A = 210$ to ~ 240 within the framework of generator-coordinate extension of the Hartree–Fock–Bogoliubov self-consistent mean field theory, using Gogny interactions. Figure 4.25 a plot of the systematics of predicted $B(E3 : 0^+ \rightarrow 3^-)$ values in these nuclei. These theoretical results are very interesting in that they predict enhanced $B(E3 : 0^+ \rightarrow 3^-)$ values in several of these nuclei. An experimental verification of these predictions is required that will point to stable octupole deformation in these nuclei. However, as yet only in a few of these cases, the octupole moments could be measured as these are very difficult measurements.

Figure 4.26 gives a comparison of experimentally derived values of the intrinsic quadrupole moment, Q_2 (also the symbol Q_0 has been used in this book) and the intrinsic octupole moment, Q_3 for the ^{208}Pb, ^{220}Rn, ^{224}Ra, ^{226}Ra, ^{230}Th, ^{232}Th and ^{234}U even-even nuclei as a function of mass number A. As mentioned in the figure caption, the values of Q_2 and Q_3 are from Table 2 in [41]. These values were obtained from the relevant measured matrix elements obtained in Coulomb excitation experiments. We follow the discussion given in [74]. It is found that there is a large ($\sim$ factor of 6) change in $E2$ moment from ^{208}Pb to ^{234}U which is not so for the $E3$

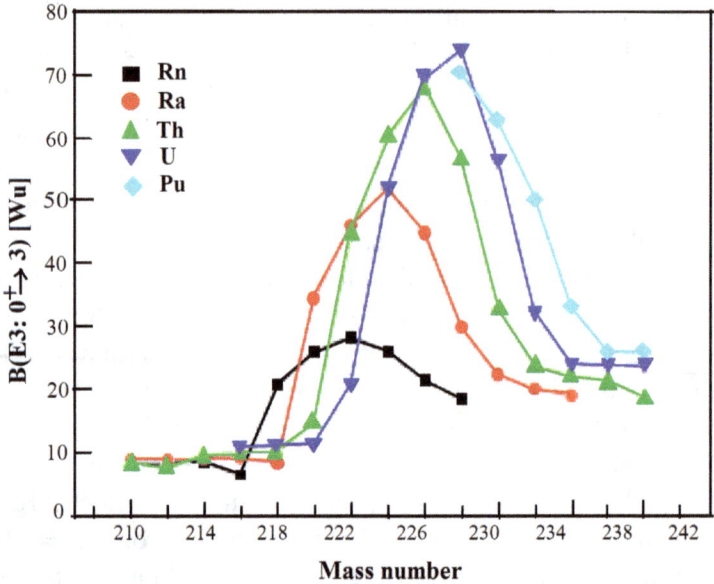

Fig. 4.25. Systematics of theoretical values of $B(E3 : 0^+ \to 3^-)$ W.u. [73] for the even-even Rn, Ra, Th, U and Pu nuclei as a function of mass number A. The original figure is courtesy of Professor P.A. Butler.

moments. The $E3$ moments for ^{208}Pb, ^{220}Rn, ^{230}Th, ^{232}Th and ^{234}U are similar, consistent with these being octupole vibrators. On the other hand, the nuclei ^{224}Ra and ^{226}Ra exhibit larger $E3$ moments which signal enhanced octupole collectivity. This is consistent with the onset of octupole deformation in this mass region. It will, however be interesting to see, if in the future, the Q_3 moments in some of the even-even ^{222}Ra and ^{228}Ra and the Th nuclei near $A = 224$ (especially ^{224}Th and ^{226}Th) could be experimentally determined to learn more about the systematic behavior of octupole moments in this mass region of the actinide nuclei (see **Note added in Proof** on p. 152).

Let us now consider nuclei in the lanthanide region. Figure 4.27 shows the systematics of experimentally measured values of $E3$ moments in even-even nuclei from Xe to Dy in $N = 82$ to 90 region. In the recent measurements of $B(E3 : 3^- \to 0^+)$ in Coulomb excitation experiments in $N = 88$ ^{144}Ba [75] and the $N = 90$ ^{146}Ba [65] nuclei,

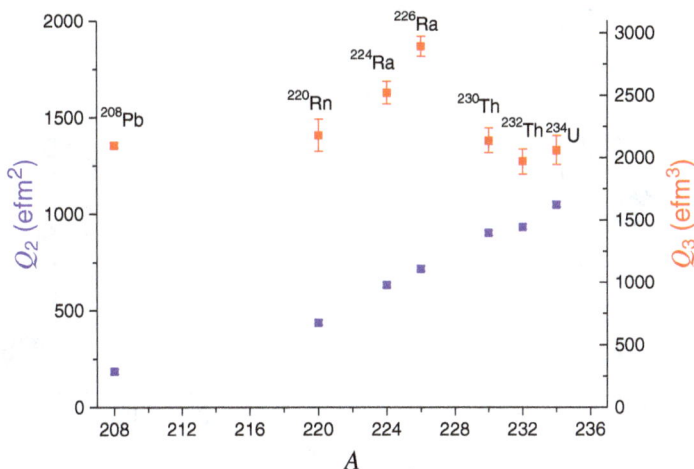

Fig. 4.26. Plot of experimentally obtained intrinsic quadrupole moment Q_2 (also denoted as Q_o in this book) ($\bullet$) (left vertical scale) and intrinsic octupole moment Q_3 ($\blacksquare$) (right vertical scale) versus mass number A, in ^{208}Pb, ^{220}Rn, ^{224}Ra, ^{226}Ra, ^{230}Th, ^{232}Th and ^{234}U even-even nuclei. For the experimental values of Q_2 and Q_3 used in this figure, see Table 2 in [41]. (See also Table 4.3 for Q_o values adopted in Ra and Th isotopes in this book.) The figure is courtesy of Professor P.A. Butler.

the $B(E3 : 3^- \rightarrow 0^+) = 48 \, (^{+25} \, _{-34})$ W.u. and $48 \, (^{+21} \, _{-29})$ W.u., respectively, were obtained. The results of both the above experiments give direct evidence of enhanced octupole collectivity. The analysis in ^{144}Ba gave a value $Q_3 = 1.73 \, (^{+45} \, _{-62}) \times 10^3 \, e \, \text{fm}^3$. The value of $E3$ moment in ^{146}Ba, since it is not available, is not shown in the figure. For a discussion of the results via various theoretical predictions on both these Ba nuclei, see the above two references and [66]. The largest theoretical value of $B(E3 : 3^- \rightarrow 0^+) = 25$ W.u. for ^{144}Ba is obtained in [66]. This value, even if low in comparison to the experimental value, it is in agreement within the large experimental error.

The experimental determination of octupole moments in ^{224}Ra, ^{226}Ra and ^{144}Ba, ^{146}Ba tell us that there is enhanced octupole collectivity in these nuclei but it does not differentiate between dynamic and static octupole collectivity. In order to distinguish between these two modes, additional or supplementary experimental

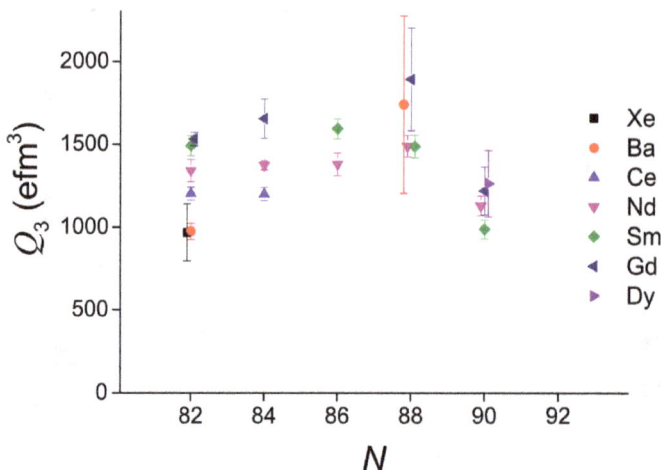

Fig. 4.27. Plot of experimentally obtained intrinsic octupole moment Q_3 in the even-even Xe to Dy nuclei versus neutron number N. For the experimental data, see [71, 75]. The figure is courtesy of Professor P.A. Butler.

information is needed. For this purpose, let us look for it through the behavior of energy levels in some of these nuclei.

Let us first consider the actinides. The energy level systematics in the positive parity ground state bands and in the negative parity bands in the even-even Ra and Th nuclei are shown in Fig. 3.1. See also a similar plot (Fig. 4.22), exclusively for the excitation energies of the 3^- states in these nuclei. These excitation energy plots clearly indicate that for the $N = 134, 136$ and 138 nuclei, since the excitation energy of the negative parity states (specifically, say for the 3^- states are lowest), the octupole collectivity is expected to be the strongest. This is in agreement with the results obtained from the intrinsic octupole moment Q_3 for ^{224}Ra and ^{226}Ra.

Let us talk a bit about the determination of stable shape of a nucleus. In the well deformed nuclei in the rare-earth mass region $150 < A < 190$, in the yrast bands, the $B(E2; I \rightarrow I - 2)$ values as a function of spin, before the band-crossing region, generally obey the rigid rotor model although there are some deviations found recently in ^{168}Yb nucleus [76]. The $B(E2; I \rightarrow I - 2)$ values within the band can be related to the intrinsic quadrupole moment Q_0 by the rigid rotor model formula (see Eq. (4.2)). To investigate the behavior of transition strengths within a band, the above-mentioned rigid rotor formula may be modified by replacing Q_o by Q_t, the transition quadrupole moment to characterize each $I \rightarrow I - 2$ transition — Q_t = e.Q_o for a $K = 0$ band [21]. For a nucleus where the nuclear shape is not affected by rotation at low spins, $Q_t (I \rightarrow I - 2)/Q_t(2^+ \rightarrow 0^+)$ should be constant for all the band members before the backbending region. (see e.g. Fig. 9 in [76]). If so, then this determines the stable quadrupole deformed shape for the nucleus.

One can use the above logic also for the octupole deformed nuclei. However, in these nuclei, in Coulomb excitation experiments, it is difficult to measure the $E2$ and $E3$ matrix elements as a function of spin. Success was achieved, however, only in the case of ^{226}Ra [77], in the actinide region. The data, in this nucleus could be fitted to a constant intrinsic quadrupole and octupole moments. In ^{224}Ra, data could only be obtained to perform such a fit for the constant intrinsic quadrupole moment [41].

Let us now discuss the nuclear shape information in ^{144}Ba and ^{146}Ba nuclei. The behavior of the observed energy levels and the gamma-ray transitions between them can often be an indicator for such information. The level schemes of ^{144}Ba and ^{146}Ba are shown in Figs. 4.28 and 4.29 respectively. In both the nuclei enhanced $E1$ transitions are observed from the positive parity levels to the negative parity levels and also vice versa, between Band A and Band B. However, in ^{146}Ba fewer $E1$ transitions have been found. In comparison to ^{144}Ba, in ^{146}Ba the intrinsic electric dipole moment D_o is small due to the cancellation of the contributions to D_o from the neutron and the proton shell correction terms (see Sec. 4.4 for details). The positive and the negative parity bands in these two nuclei become nearly interleaved after spin ~9 ℏ (see Fig. 3.7). The displacement energy $\delta E\ (I)$, in these nuclei, approaches a value of zero at the above-mentioned spin (see Fig. 3.10, upper panel and Table 3.1). This value is expected for a nucleus when it assumes the stable octupole deformed shape. As mentioned above (see Fig. 4.23), the energy of the 3^- states in ^{144}Ba and ^{146}Ba are lowest amongst

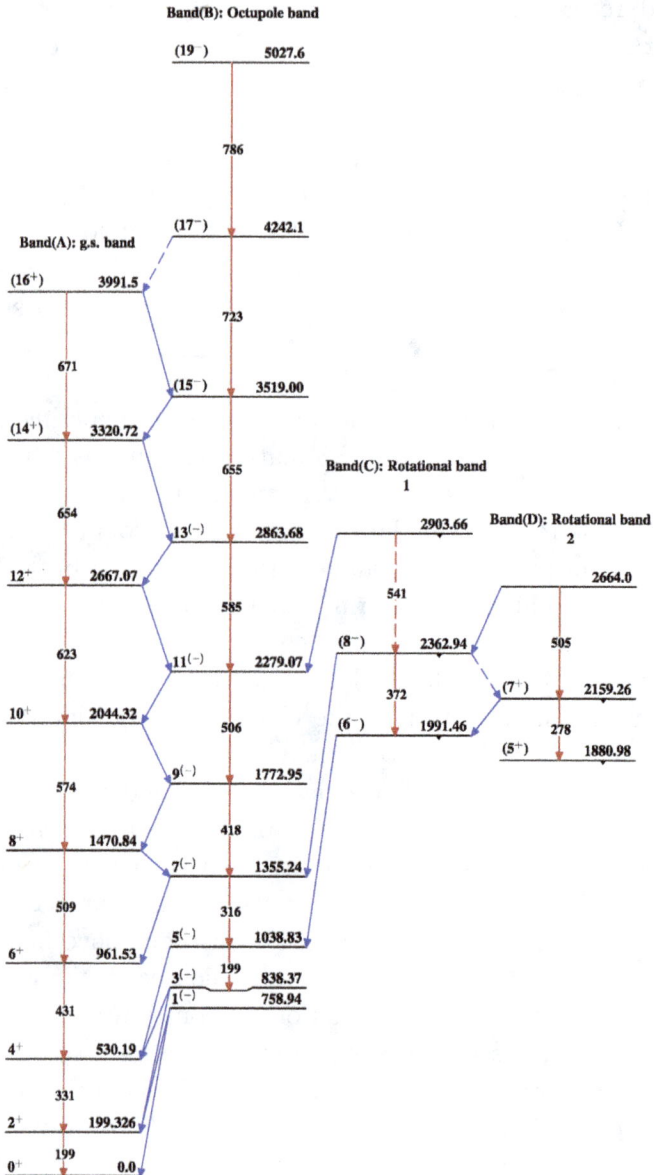

Fig. 4.28.   Level scheme of ^{144}Ba as given in [3 and references therein].

Band(C): Side band 1

(17⁻) 4071.9

Band(A): g.s. band

(16⁺) 3737.2

620

545

Band(D): Side band 2

(15⁻) 3452.39

14⁺ 3192.66

514

(14⁻) 3297.7

Band(E): Side band 3
based on 1974 level

560

Band(B): Octupole band

2953.5

507

Band(F): Side band 4
based on 2097 (7⁻)
level

13⁻ 2876.44

13⁻ 2938.74

12⁺ 2632.31

584

423

(12⁻) 2790.83

424

2710.2

11⁻ 2516.05

(10,11⁻) 2530.0

360

580

11⁻ 2292.59

325

402

(10⁻) 2389.27

317

(9⁻) 2349.91

9⁻ 2191.24

299

(8,9⁻) 2213.04

253

10⁺ 2052.01

515

246

8⁻ 2090.46

(7⁻) 2096.89

7⁻ 1944.77

216

1974.4

6⁻ 1874.72

1777.60

569

429

8⁺ 1482.63

7⁻ 1349.06

524

325

5⁻ 1024.53

6⁺ 958.37

203

3⁻ 821.10

445

1⁻ 82 738.82

4⁺ 513.66

332

2⁺ 181.04

181

0⁺ 0.0

$^{146}_{56}Ba_{90}$

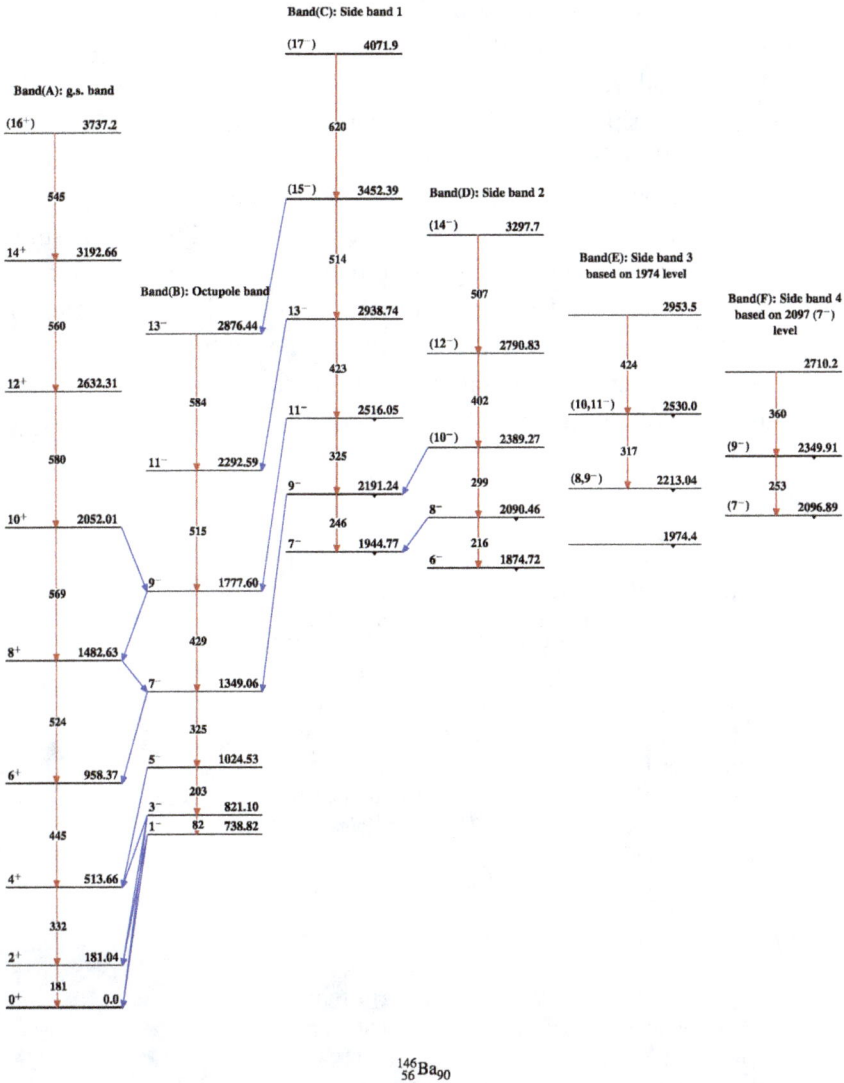

Fig. 4.29. Level scheme of ^{144}Ba as given in [3 and references therein]. See also Fig. 4.15.

the other neighboring Ba and the Ce and Nd nuclei. A plot of the difference in the aligned angular momentun $\Delta i_x = i(-) - i(+)$ versus rotational frequency $\hbar\omega$ in the even-even Ba nuclei (Fig. 3.23), however, shows that in the mid-frequency region, ^{144}Ba and ^{146}Ba exhibit octupole vibrational nature.

In conclusion, some experimental observables from the energy levels in ^{144}Ba and ^{146}Ba do indicate that these nuclei tend to assume a stable octupole deformed shape with increase in spin or rotation. In this situation, for direct evidence on stable octupole deformation in these nuclei, the experimentally obtained enhanced values of $B(E3 : 3- \rightarrow 0^+)$ and octupole moment Q_3, from the Coulomb excitation experiments [65, 75] are found to be consistent.

References

1. J.R. Hughes *et al.*, *Nucl. Phys. A* **512**, 275 (1990).
2. M. Gai *et al.*, *Phys. Lett.* **215**, 242 (1988).
3. Brookhaven National Data Center, ENSDF files: http://www.nndc.bnl.gov.
4. Y. Gono *et al.*, *Nucl. Phys. A* **459**, 427 (1986).
5. N. Schulz *et al.*, *Phys. Rev. Lett.* **63**, 2645 (1989).
6. J.F. Smith *et al.*, *Phys. Rev. Lett.* **75**, 1050 (1995).
7. J.F.C. Cocks *et al.*, *Nucl. Phys. A* **645**, 61 (1999).
8. A. Lopez-Martens *et al.*, *Eur. Phys. J. A* **50**, 132 (2014).
9. W. Reviol *et al.*, *Phys. Rev. C* **74**, 044305 (2006).
10. S. Raman *et al.*, *At. Data Nucl. Data Tables* **78**, 1 (2001).
11. M. Wieland *et al.*, *Phys. Rev. C* **45**, 1035 (1992).
12. D. Ward *et al.*, *Nucl. Phys. A* **406**, 591 (1983).
13. B. Ackermann *et al.*, *Nucl. Phys. A* **559**, 61 (1993).
14. W. Urban *et al.*, *Acta Phys. Pol. B* **32**, 2527 (2001).
15. E. Garrote *et al.*, *Phys. Rev. Lett.* **80**, 4398 (1998).
16. W. Reviol *et al.*, *Phys. Rev. C* **90**, 044318 (2014).
17. G. Maquart *et al.*, *Phys. Rev. C* **95**, 034304 (2017).
18. W. Urban *et al.*, *Nucl. Phys. A* **613**, 107 (1997).
19. W.R. Phillips *et al.*, *Phys. Lett. B* **212**, 402 (1988).
20. W. Urban *et al.*, *Phys. Lett. B* **200**, 424 (1988).
21. *Gamma-ray and Electron Spectroscopy in Nuclear* Physics by H. Ejiri and M.J.A. de Voigt, Clarendon Press, Oxford (1989) p. 503–504.
22. B. Pritychenko *et al.*, *At. Data and Nucl. Data Tables* **107**, 1 (2016).
23. *In-Beam Gamma-ray Spectroscopy* by H. Morinaga and T. Yamazaki, (North-Holland Publishing Co., 1976) p. 70.
24. K.E.G. Löbner in *The Electromagnetic Interaction in Nuclear Spectroscopy*, (North Holland Publishing Co., 1975) p. 149.

25. T.K. Alexander and J.S. Forster in *Adv. Nucl. Phys.* **10**, (1978) p. 201.
26. C.F. Perdrisat, *Rev. Mod. Phys.* **38**, 41 (1966).
27. P. Zeyen *et al.*, *Z. Phys. A* **328**, 399 (1987).
28. W.R. Phillips *et al.*, *Phys. Rev. Lett.* **57**, 3257 (1986).
29. A. Bohr and B.R. Mottelson, *Nuclear Structure*, Vol. 2 (Benjamin, New York, 1975).
30. M. Dahlinger *et al.*, *Nucl. Phys. A* **484**, 337 (1988).
31. P.A. Butler and W. Nazarewicz, *Nucl. Phys. A* **533**, 249 (1991).
32. P.A. Butler and W. Nazarewicz, *Rev. Mod. Phys.* **68**, 349 (1996).
33. N.J. Hammond *et al.*, *Phys. Rev. C* **65**, 064315 (2002).
34. L.M. Robledo and G.F. Bertsch, *Phys. Rev. C* **86**, 054306 (2012).
35. L.M. Robledo, *Eur. Phys. J. A* **52**, 300 (2016).
36. L. Grodzins, *Phys. Lett.* **2**, 88 (1962).
37. T. Ishii *et al.*, *Nucl. Phys. A* **444**, 237 (1985).
38. S. Raman *et al.*, *Atomic Data and Nucl. Data Tables* **36**, 1 (1987).
39. H. Mach *et al.*, *Eur. Phys. J. A* **52**, 172 (2016).
40. L.P. Gaffney, Ph.D., Thesis, University of Liverpool (2012).
41. L.P. Gaffney *et al.*, *Nature* **497**, 199 (2013).
42. S.K. Tandel *et al.*, *Phys. Rev. C* **87**, 034319 (2013).
43. W. Urban *et al.*, *Phys. Lett. B* **247**, 238 (1990).
44. Priv. Comm. B. Bucher in e-mail of May 2017.
45. H. Mach *et al.*, *Phys. Rev. C* **41**, R2469 (1990).
46. S.J. Zhu *et al.*, *Phys. Lett. B* **357**, 273 (1995).
47. D.C. Biswas *et al.*, *Phys. Rev. C* **71**, 011301 (2005).
48. V.M. Strutinski, *Atomnaya Energiya* **4**, 150 (1956).
49. V.M. Strutinski, *J. Nucl. Energy* **4**, 523 (1957).
50. A. Bohr and B.R. Mottelson, *Nucl. Phys.* **4**, 529 (1957).
51. A. Bohr and B.R. Mottelson, *Nucl. Phys.* **9**, 687 (1958/59).
52. G.A. Leander, AIP Conf. Proc. 125 (American Institute of Physics, New York, 1985) p. 125.
53. G.A. Leander, *Nuclear Structure '85*, Proc. Niels Bohr Cent. Conf., Copenhagen, 1985, eds. R.A. Broglia *et al.*, (North Holland 1985) p. 249.
54. G.A. Leander *et al.*, *Nucl. Phys. A* **453**, 58 (1986).
55. W. Nazarewicz, *Nucl. Phys. A* **520**, 333c (1990).
56. C.O. Dorso *et al.*, *Nucl. Phys. A* **451**, 189 (1986).
57. I. Hamamoto *et al.*, *Phys. Lett.* **226**, 17 (1989).
58. J.L. Egido and L.M. Robledo, *Nucl. Phys. A* **494**, 85 (1989).
59. J.L. Egido and L.M. Robledo, *Nucl. Phys. A* **518**, 475 (1990).
60. J.L. Egido and L.M. Robledo, *Nucl. Phys. A* **545**, 589 (1992).
61. A. Tsvetkov *et al.*, *J. Phys. G: Nucl. Part. Phys.* **28**, 2187 (2002).
62. T.M. Shneidman *et al.*, *Phys. Rev. C* **67**, 014313 (2003).
63. L.M. Robledo *et al.*, *Phys. Rev. C* **81**, 034315 (2010).
64. Y.Yu. Denisov, *Eur. Phys. J. A* **47**, 80 (2011).
65. B. Bucher *et al.*, *Phys. Rev. Lett.* **118**, 152504 (2017).
66. R.N. Bernard *et al.*, *Phys. Rev. C* **93**, 061302(R) (2016).

67. P.D. Cottle and D.A. Bromley, *Phys. Lett. B* **182**, 129 (1986).
68. R.K. Sheline and M.A. Riley, *Phys. Rev. C* **61**, 057301 (2000).
69. P.D. Cottle, *Phys. Rev. C* **42**, 1264 (1990).
70. M. Spieker *et al.*, *Phys. Rev. C* **88**, 041303(R) (2013).
71. T. Kibédi and R.H. Spear, *At. Data and Nucl. Data Tables* **80**, 35 (2002).
72. R.H. Spear and W.N. Catford, *Phys. Rev. C* **41**, R1351 (1990).
73. L.M. Robledo and G.F. Bertsch, *Phys. Rev. C* **84**, 054302 (2011).
74. P.A. Butler, *J. Phys. G: Nucl. Part. Phys.* **43**, 073002 (2016).
75. B. Bucher *et al.*, *Phys. Rev. Lett.* **116**, 112503 (2016).
76. P. Petkov *et al.*, *Nucl. Phys. A* **957**, 240 (2017).
77. H.J. Wollersheim *et al.*, *Nucl. Phys. A* **556**, 261 (1993).

Note added in proof: In a recent multistep Coulomb excitation experiment (P.A. Butler *et al. Phys. Rev. Lett.* **124**, 042503 (2020)), the observed pattern of electric octupole (E3) matrix elements was explained by describing ^{222}Ra with stable octupole deformation (pear-shape) and ^{228}Ra as octupole vibrator.

Chapter 5

Status, Conclusions and Perspectives

5.1. Introduction

In Sec. 5.2 of this chapter, a summary and review of the experimental findings are presented. Since a large number of experimental investigations have been carried out on nuclei exhibiting octupole correlations, in the actinide and the lanthanide nuclei, it is very useful for the readers to acquaint with the salient findings. Detailed account of the investigations are mentioned in Chapters 1–5. Theoretical interpretation of experiment data and theoretical predictions for the octupole shaped nuclei have been carried out in a large number of such investigations as mentioned in literature. These have proved to be of immense value in the understanding of the static octupole deformation, the octupole vibrational phenomena and nuclear structure, in such nuclei.

The matter-antimatter asymmetry problem [1, 2] in the Universe is very interesting but not understood yet. Within the framework of the Standard Model (SM) of Particle Physics, it is related to the Baryon asymmetry (Baryon-antibaryon asymmetry) problem. The measured Baryon asymmetry is $(6.1 \pm 0.3) \times 10^{-10}$ whereas that from the SM is about 10^{-18}. The charge-parity (CP) violating interactions in matter may be a source of this asymmetry. Parity violation has been found in beta decay [3, 4], CP violation in K^0 meson decays [5], CP violation in B^o meson decays [6–8] and the latest discovery on CP violation in D^0 meson decays [9]. The topic of CP violation has been extensively discussed in literature in research articles and books. The CP violating interactions may also manifest nonzero electric dipole moment (EDM) in particles, molecules and atomic nuclei. In Sec. 5.3, we discuss the atomic EDMs in pear-shaped odd-A high-Z light actinide nuclei where the EDM is theoretically predicted to be highly enhanced. The pear-shaped nuclei, therefore, are considered important in very sensitive experimental search for CP-violating interactions.

5.2. Summary and Review of Experimental Findings

5.2.1. *Energy levels*

5.2.1.1. *Even-even nuclei*

About six decades back, in a series of pioneering experiments, negative parity 1^- states in addition to the positive parity rotational bands were found in the quadrupole deformed even-even nuclei in the light actinide region. This observation was attributed to the nuclei having a pear-shape (Chapter 1) and it marked the beginning of investigations, both experimental and theoretical, in such nuclei.

Later, evidence of quadrupole-octupole shapes in light actinide nuclei was also found from spin and parity J^π of the ground states of these nuclei. The experimentally determined J^π values were not consistent with the Nilsson model predictions. However, these were later found to be consistent with model calculations when taking the quadrupole-octupole nuclear shapes into consideration (see Sec. 2.4).

In the present section, a summary and review of experimental findings from various investigations that were done in the light actinide and the lanthanide nuclei where octupole correlations have been found, will be given. Most of these nuclei lie in the transitional region between the quadrupole vibrators and the deformed quadrupole rotors (see Fig. 2.1).

In recent years, a number of theoretical calculations on the evolution of octupole shapes in the ground state of the light actinide and the lanthanide nuclei were undertaken. In very recent calculations, with the details discussed in Sec. 2.3, the ground state quadrupole-octupole shape evolution was found in several of these nuclei. A beautiful illustration of these shapes are given in the diagrams for potential energy surfaces in the (β_2, β_3) plane (see Figs. 2.2–2.4). For details of shape predictions and their comparison with experimental data, see Sec. 2.3 and Figs. 2.5, and 2.6.

During the last about four decades, several experimental high spin spectroscopic investigations have been done, as cited in literature, in the light actinide and the lanthanide nuclei, and the emerging energy level patterns have been determined. A representative level scheme of ^{222}Th is shown in Fig. 5.1. One of the most characteristic features or signature of reflection-asymmetric shape in an even-even nucleus is the observation of close-lying even-spin positive parity $J^\pi = 0^+, 2^+, 4^+, \ldots$ band and an odd-spin negative parity $J^\pi = 1^-, 3^-, 5^- \ldots$ band. Many of the neighboring opposite parity states in these alternating parity bands are connected by enhanced E1 transitions. The comparison of excitation energy systematics of positive and the negative parity band levels in the even-even Ra and Th isotopes (Fig. 3.1) as a function of neutron number, depicts for the negative parity levels, a peculiar parabolic behavior with minima for isotopes in the neutron number range $N = 134$–138, signifying the region of stable octupole deformation. In the even-even Ba isotopes, the negative parity states do not behave in the way the Ra and Th nuclei do but for $N = 88$–92 region, the negative parity states are the lowest in excitation energy (Sec. 3.2, Fig. 3.2, top panel). Considering energy splitting in these alternating parity bands, it is found that above $J \sim 7$ ℏ (in light actinides), it nearly vanishes and the positive and negative parity states form interleaved alternating parity $J = 0^+, 1^-, 2^+, 3^-, 4^+, 5^-, \ldots$ band. The situation is different for heavier isotopes of Ra and Th: in ^{226}Ra $(N = 138)$ and ^{228}Ra $(N = 140)$, the energy difference at low spins tends to disappear only at higher spins of $J \sim 11$ ℏ and 18 ℏ, respectively. In Sec. 3.3, the energy splitting situation is considered in detail as a function of angular momentum in even-even Ra, Th and Ba, Sm nuclei (see also Figs. 3.3–3.8). Within the framework of reflection asymmetric shell model, the energy of the states in the even-even Ra isotopes was calculated in the positive and the negative parity bands as a function of spin and compared with experimental data. The agreement between experiment and theory is found to be excellent.

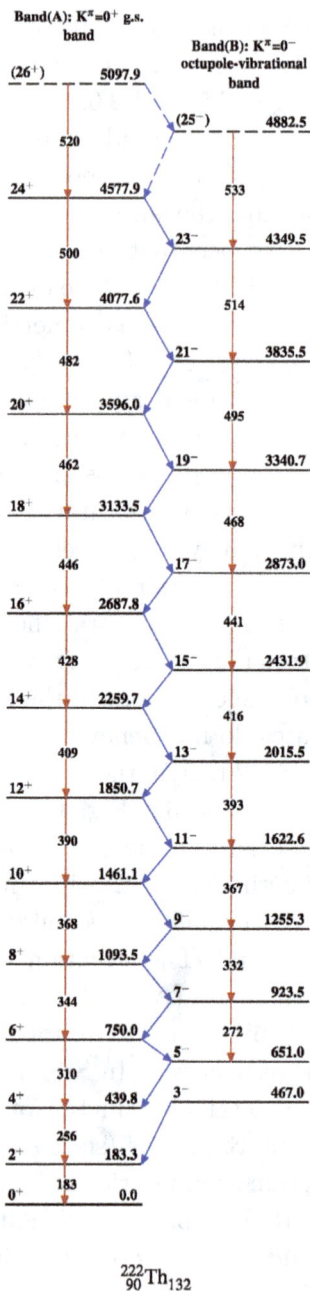

Fig. 5.1.　Level scheme of ^{222}Th [10].

In order to further understand the characteristics of reflection asymmetric even-even nuclei, the energy displacement (displacement energy, also called parity splitting) between the positive and the negative parity bands, was defined. The energy difference between a state at a particular odd spin I in the negative parity band and a state at the same odd spin I in the positive parity band can be found in a simple manner by subtracting from the energy of the negative parity odd spin state I, a value calculated by taking an average of the energies of the adjacent even spin states in the positive parity band (see Sec. 3.4). The displacement energies $\delta E(I)$ were calculated in a number of even-even nuclei in Ra, Th, Ba, Sm and the $N = 88$ isotones in the $A \sim 150$ mass region and plots of these versus spin are shown in Figs. 3.9–3.11. There is one observation which appears to be a common general feature, and it is that at low spins, the displacement energies decrease or approach the line with $\delta E(I) = 0$, as a function of the increase in spin. This zero value of displacement energy is expected for stable octupole deformation. In ^{220}Ra, ^{222}Ra, ^{224}Ra, ^{222}Th, ^{224}Th, ^{142}Ba, ^{144}Ba,^{146}Ba, ^{146}Ce and ^{148}Nd nuclei, $\delta E(I) = 0$ is achieved at spin $I\sim9$ ħ.

One question of significance in nuclei exhibiting octupole correlations is how to distinguish between the octupole vibrational and octupole deformed nuclei using the data on energy levels. With this aim in mind, in Sec. 3.5, the ratio of rotational frequencies in the negative and the positive parity bands $\omega_{\text{rot}}(\pi = -)/\omega_{\text{rot}}(\pi = +)$ and staggering in the energy ratio $E(J^\pi)/E(2^+)$ between the positive and the negative parity band levels, as a function of spin, were examined. Also, the difference $\Delta i_x = [i(-) - i(+)]$ between particle aligned angular momentum i between the negative parity and the positive parity bands, as a function of rotational frequency, was considered.

The ratio of rotational frequencies is equal to unity for a perfectly reflection asymmetric nucleus. In another limit, this ratio should be equal to $[4(I - 3) - 2]/(4I - 2)$ for the rotation of an aligned octupole phonon. These frequency ratios, as a function of spin, were considered in the even-even isotopes of Ra, Th, Ba, Sm and the $N = 88$ isotones (Sec. 3.5.1). ^{222}Ra ($N = 134$), ^{224}Ra ($N = 136$) and ^{224}Th ($N = 134$)

nuclei attain the limit for the stable octupole deformation earliest at spin $I \sim 12\,\hbar$ (Fig. 3.12), whereas the nuclei ^{230}Th $(N = 140)$ and ^{232}Th $(N = 142)$ depict the octupole vibrational trend. In the isotopes ^{142}Ba, ^{144}Ba, ^{146}Ba and ^{148}Ba with N from 86 to 92, the ratio approaches the limit for stable octupole deformation at $I \sim 12 - 14\,\hbar$ (Fig. 3.13, upper panel). The nucleus ^{152}Sm $(N = 90)$ follows the aligned octupole vibrational trend (Fig. 3.13).

In Sec. 3.5.2, plots of experimental excitation energy ratio $E(J^\pi)/E(2^+)$ for each $J^\pi (= 1^-, 2^+, 3^-, 4^+, \ldots)$ state in the positive and the negative parity bands to that of the 2^+ first excited state, as a function of spin I, for the even-even nuclei in Ra, Th, Ba and the $N = 88$ isotones, were considered (see Figs. 3.15–3.18). One finds that the amplitude of the $\Delta I = 1$ energy ratio staggering between the adjacent negative and the positive parity levels, at low spins, (a) increases from lighter to heavier mass nuclei, i.e., with the increase in neutron number N, (b) in the lighter mass nuclei this energy ratio staggering is almost negligible and (c) for a nucleus, it gradually decreases with the increase in spin. After a certain high spin I, in general, the two rotational bands merge into a single interleaved alternating parity band. This phenomenon seems to be common for all the nuclei considered. The negligible $\Delta I = 1$ energy ratio straggling is a signature for the nucleus to attain stabilization of octupole deformed shape with an increase in rotation. A number of theoretical investigations on the $E(J^\pi)/E(2^+)$ excitation energy ratio straggling as a function of spin, details of which are given in Sec. 3.5.2, have been carried out in many of the even-even actinide and lanthanide nuclei. The theoretical study in ^{224}Ra showed that the spin dependent odd-even energy ratio staggering at low spins is related to the rotation induced octupole shape stabilization of the positive parity states due to the gradual drift (increase) of their octupole shape with increasing spin towards the octupole shape of the negative parity states which stays almost constant with the increase in spin. In the plot of experimental excitation energy ratio $E(J^\pi)/E(2^+)$ for ^{224}Ra (see Fig. 3.15), the excitation energy ratio staggering diminishes with the increase in spin at low spins. At spin $I \sim 8\,\hbar$, the straggling becomes small.

This result is in agreement with the theoretical predictions. In another theoretical study on even-even Ba isotopes, it was found that (i) the calculated straggling in the energy ratio $E(J^\pi)/E(2^+)$ as a function of spin I, except in the case of ^{142}Ba, was found to be in agreement with the experimental energy ratios, (ii) in ^{144}Ba nucleus, for different combinations of the values of β_2 and β_3, e.g., for $\beta_3 = 0.1$, the straggling amplitude roughly increases with the increase in the quadrupole deformation parameter $\beta_2(= 0.0, 0.1, 0.2, 0.3, 0.4)$. This was also found experimentally in the Ra and Th isotopes, i.e., increase in straggling amplitude in nuclei from lighter to the heavier ones and (iii) the octupole shape of the negative parity states, with increase in spin remains almost stable whereas for the positive parity states, it drifts from weakly octupole to that of the negative parity states at higher spins, as in the case of ^{224}Ra. The rotation induced octupole shape stabilization in the positive parity states can then be inferred as a common phenomenon in all the actinide and lanthanide nuclei under discussion. On the basis of the theoretical and experimental investigations of $E(J^\pi)/E(2^+)$ excitation energy ratio, straggling as a function of spin, an indication, in general, can be derived that the nuclei exhibiting low or negligible straggling from low spins up to the high spins are octupole deformed.

Now, let us try to look into the information that can be obtained from the behavior of the difference quantity Δi_x of aligned angular momentum i (or i_x) between the positive and the negative parity bands, as a function of rotational frequency $\hbar\omega$. One situation is when the component of the angular momentum aligned to the rotation axis for the negative parity state $i(-)$ or for the positive parity state $i(+)$, at the same rotational frequency, is equal to the rotational angular momentum R. Then, in this case $\Delta i_x = [i(-) - i(+)] = 0$. This will happen for a permanent octupole deformed nucleus. The other situation could be when the negative parity state is formed by octupole vibrations of the rotating quadrupole deformed nucleus by the coupling of the rotational angular momentum R and the angular momentum of the octupole phonon ($3\,\hbar$). If the phonon angular momentum ($3\,\hbar$) is aligned with the rotational angular momentum R, then the quantity $\Delta i_x = 3\hbar$. This will be the case when the nucleus is octupole vibrational. Plots of $\Delta i_x = [i(-) - i(+)]$, as a function of rotational frequency $\hbar\omega$, for the even-even ^{218,220,222}Rn ($N = 132, 134$ and 136), $^{220, 222, 224, 226, 228}$Ra ($N = 132$ to 140) and ^{222,224,226,228,230}Th isotopes ($N = 132$ to 140) are shown in Fig. 3.22 (Sec. 3.5.3). All the three Rn isotopes considered behave like octupole vibrators ($\Delta i_x \sim 3\,\hbar$) in almost the entire rotational frequency range. In the three isotopes of Ra, $^{222, 224, 226}$Ra with $N = 134$, 136 and 138 and the isotopes ^{224}Th and ^{226}Th with $N = 134$ and 136 respectively, octupole deformation stabilizes $\{\Delta i_x \rightarrow 0\}$ at high rotational frequencies ($\hbar\omega \sim 0.20$ MeV). Figure. 3.23 (Sec.3.5.3), shows plots similar to those in Fig. 3.22, for the even-even Xe, Ba, Ce, Nd and the Sm nuclei. From these plots, it is apparent that most of these nuclei in this region behave to qualify as octupole vibrators in a large part of the rotational frequency region. However, at high rotational frequencies, $N = 88$ ^{144}Ba, and the $N = 90$ isotones, ^{148}Ce and ^{150}Nd exhibit trends of being octupole deformed.

The rotational properties of reflection asymmetric nuclei in the region of light even-even actinides and the lanthanides are summarized below (see Sec. 3.6 for details). For this purpose, plots of total aligned angular momentum I_x and kinematic moment of inertia, $\mathcal{J}^{(1)}$ as a function of rotational frequency $\hbar\omega$, are considered. The behavior of the difference in aligned angular momentum Δi_x has already been discussed above. In the $N = 130$ isotones ^{218}Ra and ^{220}Th an irregular zig-zag behavior around nearly a constant rotational frequency is exhibited in plots of total aligned angular momentum I_x as a function of rotational frequency (see Fig. 3.24). This indicates that these nuclei seem to be soft quadrupole vibrators. The $N = 132$ isotones ^{220}Ra and ^{222}Th exhibit a contrasting behavior (see level schemes in Fig. 3.25). In ^{220}Ra, the positive parity band is seen up to $J^\pi = 28^+$ (30^+) and the negative parity band up to $J^\pi = 29^-$ (31^-) whereas in ^{222}Th these bands are observed only up to $J^\pi = 24^+$ and $J^\pi = 23^-$ (25^-). In the plots of total aligned angular momentum, I_x as a function of rotational frequency, $\hbar\omega$, for the yrast positive parity ground state bands and the negative parity octupole bands in these two isotones (Fig. 3.26), a rather regular increase in I_x with rotational frequency $\hbar\omega$ is found, indicating quadrupole collectivity. In plots of kinematic moment of inertia, $\mathcal{J}^{(1)}$, as a function of rotational frequency, $\hbar\omega$, in these $N = 132$ isotones ^{220}Ra and ^{222}Th (Fig. 3.27), bandcrossings are indicated in the negative parity bands at $\hbar\omega \sim 0.21$ MeV which may be due to the ν $(j_{15/2})^2$ neutron pair aligned bands. Detailed microscopic calculations in the cranked reflection-asymmetric Woods–Saxon–Bogolyubov–Strutinsky framework were carried out by W. Nazarewicz and collaborators [11] in some of the even-even Ra and Th nuclei to study the evolution of nuclear shape with rotational frequency/angular momentum (for details see Sec. 3.6). The theoretical shape evolution predictions are confirmed by the above-mentioned experimental observations. However, theory predicts a ν $(j_{15/2})^2$ neutron pair aligned bandcrossing only in ^{222}Th. Plots of kinematic moment of inertia $\mathscr{I}^{(1)}$ as a function of rotational frequency $\hbar\omega$, for the even-even ^{222}Ra, ^{224}Ra, ^{226}Ra, ^{224}Th, ^{226}Th, ^{228}Th and ^{220}Ra, ^{222}Th nuclei (Figs. 3.28 and 3.27 respectively) exhibited a common behavior in that the kinematic moment of inertia at low rotational frequencies, in the negative parity bands, is higher than that for the positive parity bands. The kinematic moment of inertia difference between the opposite parity bands decreases with the increase of rotational frequency and then the two merge at a rotational frequency of $\hbar\omega \sim 0.2$ MeV except for ^{228}Th. This behavior is an indication of the stabilization of the octupole deformed shape as the rotational frequency increases. Several theoretical explanations have been put forward to explain the observed behavior of kinematic moment of inertia for the negative and the positive parity bands in these actinide nuclei (see Sec. 3.6). Let us now consider the behavior of kinematic moment of inertia $\mathscr{I}^{(1)}$ as a function of rotational frequency $\hbar\omega$, in the even-even ^{142}Ba$_{86}$, ^{144}Ba$_{88}$, ^{146}Ba$_{90}$ and ^{148}Ba$_{92}$ isotopes (see Fig. 3.29) and in the $N = 88$ isotones ^{144}Ba, ^{146}Ce, ^{148}Nd and ^{150}Sm (see Fig. 3.30). In all these nuclei, a common feature is the enhanced kinematic moment of inertia at low rotational frequencies in the negative parity bands in comparison to that in the positive parity bands. This property is similar to that found for the even-even actinides Ra and Th (see Figs. 3.27 and 3.28). In ^{146}Ba (see also Fig. 3.20), the ground state positive parity band exhibits an upbending at $\hbar\omega \sim 0.29$ MeV which is characteristic of band crossing by a two quasiparticle band. This is an indication of shape transition from reflection-asymmetric to reflection-symmetric in this nucleus. In ^{144}Ba (also see Fig. 3.20), the experimental data is unknown if a delayed bandcrossing occurs at $\hbar\omega \sim 0.34\,\mathrm{MeV}$.

5.2.1.2. *Odd mass nuclei*

One of the important signatures of odd-mass (odd-N and odd-Z) reflection asymmetric nuclei is the existence of parity doublet bands in the level schemes. Such bands have been found in a number of odd-N and odd-Z light actinide nuclei (see Sec. 3.7). The inter-band parity doublet states in these bands have same spin I but opposite parity. As an example of parity doublet bands, see the level scheme of ^{223}Th ($Z = 90$ and $N = 133$) Fig. 3.31 (Sec. 3.7). The parity doublet states in this nucleus are: $13/2^- - 13/2^+$, $17/2^- - 17/2^+$, ... $41/2^- - 41/2^+$, $45/2^- - 45/2^+$ in Bands 1(a) - 2(b) and $11/2^+ - 11/2^-$, $15/2^+ - 15/2^-$, ... $39/2^+ - 39/2^-$, $43/2^+ - 43/2^-$ states in 1(b) - 2(a).

The spectroscopic properties of the parity doublet bands observed in some of the odd-N nuclei are summarized below.

We first consider energy difference between the parity doublet states in odd-N, $N = 133$ isotones ^{221}Ra and ^{223}Th and $N = 135$ ^{225}Th nucleus, as a function of spin I (Fig. 3.32, Sec. 3.7). The parity doublet bands in these nuclei are nearly degenerate in energy with minimum energy difference of 2.4 keV for the $15/2^-$, $15/2^+$ pair in ^{221}Ra and maximum energy difference of 67 keV for the $11/2^-$, $11/2^+$ pair in ^{225}Th. One of the general features is the energy difference staggering between $\Delta E[\pi(+ \rightarrow -)]$ and $\Delta E[\pi(- \rightarrow +)]$ parity doublet pairs. Another is the staggering phase changes at a particular spin value, that is, there is *simplex inversion* in all the parity doublet bands in these three nuclei. No detailed theoretical explanation could be found in literature for this simplex inversion.

As discussed earlier in Sec. 3.4, for even-even nuclei, the evolution of octupole correlations can be investigated through the behavior of displacement energy (parity splitting) as a function of spin. A comparison of displacement energy in the $N = 133$ isotones ^{221}Ra, ^{223}Th and $N = 135$ ^{225}Th nucleus with those in the neighboring even-even Ra and Th nuclei, showed a common general feature. As for the neighboring even-even nuclei, parity splitting $\delta E(I)$ decreases with the increase in spin $(I - I_0)$ above the ground state spin I_0 (Fig. 3.33). Another point to be noticed is that at low spins, parity splitting in odd-N nuclei is smaller than for the neighboring even-even nuclei.

The rotational properties of the odd-N nuclei ^{223}Th and ^{225}Th were discussed in Sec. 3.9 using the plots of kinematic moment of inertia as a function of rotational frequency. A comparison of similar plots (see Figs. 3.26–3.28) for the neighboring even-even Th nuclei showed that the moment of inertia in the two odd-N nuclei, at low rotational frequencies, is larger compared to that of their even-even neighbors. This is likely due to the unpaired nucleon — a neutron. Another common feature, at low spins, is that in ^{223}Th, in the negative parity bands for both simplex bands $s = -i$ and $s = +i$, the kinematic moment of inertia is larger than that in the positive parity bands. In ^{225}Th, this is similar but not so spectacular. In both the odd nuclei, all bands attain similar values of kinematic moment of inertia at $\hbar\omega \gtrsim 0.17$ MeV. This phenomenon has also been found in the even-even Ra and Th nuclei, indicating stabilization of octupole shape with rotation.

In some of the odd mass nuclei in the actinide and the lanthanide regions, at low spins, low energy mostly magnetic dipole ($M1$) transitions with small $E2$ admixtures, are found. In Sec. 3.10, the properties associated with these $M1$ transitions were considered. The parameter $(g_K - g_R)/Q_0$ was deduced in experimental measurements for the $M1 + E2$ transitions, namely, for the $5/2^+ \rightarrow 3/2^+$ 29.86 keV, $(7/2)^+ \rightarrow 5/2^+ 31.58$ keV, $9/2^+ \rightarrow (7/2)^+ 68.74$ keV transitions in $K^\pi = 3/2^+$ band and the $(5/2)^- \rightarrow 3/2^-$ 29.60 keV, $7/2^- \rightarrow (5/2)^-$ 44.22 keV and $9/2^- \rightarrow 7/2^- 50.85$ keV transitions in the $K^\pi = 3/2^-$ band, in ^{223}Ra (see level scheme of ^{223}Ra in Fig. 3.35). The plot of $(g_K - g_R)/Q_0$ values as a function of spin (Fig. 3.36) showed these are similar and nearly constant for the two bands. This is a positive argument that these transitions originate from the pair of $K^\pi = 3/2^\pm$ bands and the nucleus ^{223}Ra is reflection asymmetric in the low spin regime. Further, in ^{223}Th nucleus (see level scheme in Fig. 3.37), the experimentally deduced $|g_K - g_R|$ parameters for the 51.3 keV $(7/2^+) \rightarrow (5/2^+)$, 93.4 keV $(11/2^+) \rightarrow$

$(9/2^+)$ $M1$ transitions in the $K^\pi = 5/2^-$ band and the 67.5 keV $(9/2^+) \rightarrow (7/2^+)$, 87 keV $(15/2^=) \rightarrow (13/2^-)$ $M1$ transitions in the $K^\pi = 5/2^+$ band were obtained. All these experimentally obtained $|g_K - g_R|$ values are plotted as a function of spin I in Fig. 3.38. It is seen from this figure that within errors the values are similar within each $K^\pi = 5/2^{+or~-}$ band and between the $K^\pi = 5/2^-$ and $K^\pi = 5/2^+$ bands. The similarity of the $|g_K - g_R|$ parameters is consistent with the interpretation that the nucleus ^{223}Th is reflection asymmetric.

5.2.2. *Electromagnetic transitions*

In Chapter 3, the properties of pear-shaped nuclei which were derived from the measurement of energy of high spin levels in such nuclei, are discussed. A summary is presented in Sec. 5.2.1. In the present section, the electromagnetic properties of inter-band $\Delta I = 1$ $E1$ transitions connecting negative (positive) parity to positive (negative) parity neighboring levels between the negative and the positive parity bands and the intra-band $\Delta I = 2$ $E2$ gamma-ray transitions between the same parity high spin levels are summarized, in the light actinide and lanthanide nuclei. We, also discuss $\Delta I = 3$ $E3$ transitions. Specifically, we discuss the $B(E1)/B(E2)$ ratios, $B(E1)$, $B(E2)$, $B(E3)$ values and the associated derived values of electric dipole moments D_0, electric quadrupole moments Q_2 and the electric octupole moments Q_3 in the reflection asymmetric nuclei. The systematic of excitation energy of 3^- states is also summarized.

The experimental $B(E1)/B(E2)$ ratio for the gamma-decay from a level with spin I is determined, from the gamma-ray branching ratio $I_\gamma(E1)/I_\gamma$ $(E2)$ and gamma-ray energies of the $E1$ and $E2$ transitions. The values of the B$(E1)$/B$(E2)$ ratios so calculated from experimental data were plotted as a function of spin in ^{218,220,222,224,226,228}Ra, ^{220,222,224,226,228}Th, ^{219}Ra, ^{221,223,225}Th isotopes and the $N = 88$ isotones ^{144}Ba, ^{146}Ce, ^{148}Nd and ^{150}Sm (see Figs. 4.1, 4.2, 4.4–4.9). The main purpose in doing so is two-fold: (i), to find if there is any angular momentum dependence of the

B(E1)/B(E2) ratios and (ii), whether there is any difference in these ratios for gamma decays from the positive and the negative parity states. In general, there was not much success in this effort, probably because of the relatively large errors in these ratios. However, in ^{218}Ra nucleus, a decrease in the *B(E1)/B(E2)* ratio with spin was found (Fig. 4.1, top panel). ^{220}Th nucleus showed a peculiar behavior, the decay from the positive parity and the negative parity states exhibited distinctly different patterns (see Fig. 4.4 (top panel) and Sec. 4.2). In the spin range 6 to $\sim$12, the ratios for the decay from the positive parity states are higher in comparison to those from the negative parity states. This can be explained in terms of energy staggering between the *E*1 transitions originating from the positive parity levels and the negative parity levels. The *E*1 transitions from the positive parity levels are favored and they compete with the branched *E*2 transitions. In ^{224}Th nucleus, an increase in the *B(E1)/B(E2)* ratio with spin was found (Fig. 4.5, top panel). Whether this effect is due to the variation of *B(E2)* with spin or due to the variation of *B(E1)*, can only be known if the absolute values of *B(E1)* and *B(E2)* are available from level lifetime measurements. The *B(E1)/B(E2)* ratios are nearly constant in ^{226}Th nucleus.

Having discussed the above B(E1)/B(E2) ratios for nuclei exhibiting octupole correlations, let us now pay attention to the absolute values of the reduced electric dipole gamma-ray transition probability *B(E1)* in even-even nuclei. The *B(E1)* values are determined from the experimentally measured *B(E1)*/B(E2) ratios and the deduced values of *B(E2)* as a function of spin using the prescription given in Sec. 4.3. The *B(E1)/B(E1)*$_w$ ratios, where *B(E1)*$_w$ are the Weisskopf estimates, were determined for a number of even-even Ra and Th isotopes and for the $N = 88$ isotones ^{144}Ba, ^{146}Ce and ^{148}Nd (see Tables 4.1 and 4.2). It is found that the E1 transitions in nuclei exhibiting octupole correlations are, in general, enhanced by about two orders of magnitude as compared to nuclei which do not show the octupole behavior.

There has been a lot of interest to interpret the experimentally obtained values of intrinsic electric dipole moments D_o and their variation with neutron number in the actinide and the lanthanide nuclei exhibiting octupole correlations.

The experimental values of intrinsic dipole moment D_0 in Ra, Th, Ba and the $N = 88$ nuclei were deduced from $B(E1)/B(E2)$ ratio and Q_0 value using the procedure as explained in Sec. 4.4. The D_0 values as a function of neutron number N are plotted in Fig. 4.11 for the Ra isotopes. The values show an increasing trend from $N = 130$ to $N \sim 132-133$ where it saturates and then it decreases with further increase in neutron number. A striking feature of this plot is the unusually low value of D_0 for $N = 136$ ^{224}Ra nucleus. A theoretical explanation of this feature was provided in terms of a macroscopic-microscopic model. In this model, the total intrinsic dipole moment is the sum of contributions from a macroscopic (liquid drop) term and the microscopic shell correction term which takes into account the effect of nuclear shell structure. It was found that in ^{224}Ra, the liquid drop and the shell correction terms mutually cancel, giving a small value for the dipole moment. It needs to be mentioned here that in $(N = 135)$ ^{223}Ra nucleus, in the $K = 1/2$ band for $I < 8$, $D_0 = 0.078 \pm 0.012$ e.fm. This is also a relatively low value of the intrinsic dipole moment. This can also be theoretically explained by the above considerations.

A plot of the intrinsic dipole moment D_0 as a function of neutron number N for the Th isotopes (Fig. 4.12) shows the increasing trend in D_0 from $N = 130$ to $N = 134$ where it saturates and then decreases with the increase in neutron number. Theory predicts the general trend found experimentally with no abnormal behavior for the Th isotopes. In one of the theoretical works, comprehensive calculations were done to obtain the macroscopic liquid drop contribution. Their results with the adopted value of the shell correction term showed good agreement for the total intrinsic dipole moment as a function of mass number for $A = 220$ to 227 Th isotopes (Fig. 4.18).

Comparing the experimentally obtained D_0 values in the ^{140}Ba, ^{142}Ba, ^{144}Ba, ^{146}Ba and ^{148}Ba nuclei (Fig. 4.14), it is seen that there is a sudden drop in the value of D_0 in ^{146}Ba $(N = 90)$ isotope. This effect is similar to that seen for the ^{224}Ra nucleus. It has also been explained within the macroscopic-microscopic model. For a more detailed theoretical approach to explain the observed low values of intrinsic dipole moments in ^{224}Ra and ^{146}Ba nuclei, see Sec. 4.4.

The excitation energy of 3^- states $E(3^-)$ in the even-even light actinide and lanthanide nuclei implies the strength of octupole collectivity in nuclei. The excitation energy systematics for the even spin positive parity and odd spin negative parity states as a function of neutron number is discussed in Sec. 3.2. Here, we exclusively consider the energy of the lowest lying 3^- states relative to the 0^+ ground state also as a function of neutron number in the even-even Rn, Ra, Th and U and Pu nuclei. Although the nature of variation of $E(3^-)$ with the neutron number is similar in both Ra-Th and U-Pu regions of nuclei but the minima of $E(3^-)$ are much shallower for the heavier nuclei in comparison to the deep minima in Ra-Th region (see Fig. 4.22). The energy of the 3^- states in ^{222}Ra, ^{224}Ra, ^{226}Ra and ^{224}Th, ^{226}Th, ^{228}Th is at the bottom of the minima. It may be mentioned here that stable octupole deformation has been found for ^{224}Ra and also for ^{226}Ra (see Sec. 4.6). The plot of energy of the lowest lying 3^- states $E(3^-)$ in the Ba, Ce and Nd even-even nuclei as a function of neutron number (Fig. 4.23) is not that spectacular as those for the actinides. The energies of 3^- states in Ba nuclei, ^{144}Ba ($N = 88$) and ^{146}Ba ($N = 90$) are the lowest. In these two nuclei strong octupole correlations have also been found (see Sec. 4.6).

In Sec. 4.6, we have considered the $E3$ single particle transition strength $|M(E3)|^2$ in Weisskopf units (W.u.) derived from experimental values of reduced electric octupole transition probabilities $B(E3$: $0^+ \rightarrow 3^-)$, as a function of neutron and proton numbers. Other quantities discussed are: theoretical predictions of $B(E3$: $3^- \rightarrow 0^+)$ values for a number of even-even heavy nuclei, like, Rn, Ra, Th, U and Pu; the experimentally derived values of the intrinsic quadrupole moment Q_2 and the intrinsic octupole moment Q_3, for the ^{208}Pb, ^{220}Rn, ^{224}Ra, ^{226}Ra, ^{230}Th, ^{232}Th and ^{234}U nuclei as a function of mass number A. Towards the end of this section, a discussion on stable octupole nuclear shapes is also given.

A plot of the $E3$ single particle transition strength $|M(E3)|^2$ versus neutron number N (Fig. 4.24, upper panel) showed enhanced octupole transition strengths or octupole collectivity at $N \sim 34$, 56, 88 and 134. This is in agreement with the lowest energies of the first 3^- states found at $N = 88$, 90 in Ba and $N = 134$, 136 in Ra, Th nuclei (see above and Sec. 4.5). The enhanced octupole transition strengths at these nucleon numbers agree with theory which predicts the likely nuclear regions for static octupole deformation.

Figure 4.25 shows the systematics of theoretically predicted B(E3 : $3^- \to 0^+$) values in Rn, Ra, Th, U and Pu nuclei as a function of mass number from $A = 210$ to $\sim$240. These theoretical results are very interesting in that they predict enhanced B($E3 : 3^- \to 0^+$) values in several of these nuclei. An experimental verification of these predictions is required which will point to stable octupole deformation in these nuclei. However, as yet only in a few of these cases, the octupole moments could be experimentally determined as these are very difficult measurements.

Figure 4.26 gives a comparison of experimentally derived values of the intrinsic quadrupole moment, Q_2 (also the symbol Q_0 has been used in this book) and the intrinsic octupole moment, Q_3 for the ^{208}Pb, ^{220}Rn, ^{224}Ra, ^{226}Ra, ^{230}Th, ^{232}Th and ^{234}U even-even nuclei as a function of mass number A. It is found that there is a large ($\sim$factor of 6) change in $E2$ moment from ^{208}Pb to ^{234}U which is not so for the $E3$ moments. The $E3$ moments for ^{208}Pb, ^{220}Rn, ^{230}Th, ^{232}Th and ^{234}U are similarly consistent as octupole vibrators. On the other hand, the nuclei ^{224}Ra and ^{226}Ra exhibit larger $E3$ moments which signal enhanced octupole collectivity. This is consistent with the onset of octupole deformation in this mass region. It will, however, be interesting to see, if in future, the Q_3 moments in some of the even-even ^{222}Ra and ^{228}Ra and the Th nuclei near $A = 224$ (especially ^{224}Th and ^{226}Th) could be experimentally determined to learn more about the systematic behavior of octupole moments in this mass region of the actinide nuclei.

The systematics of experimentally measured values of $E3$ moments in even-even nuclei from Xe to Dy in $N = 82$ to 90 region is depicted in Fig. 4.27. The recent measurements of $B(E3: 3^- \to 0^+)$ in Coulomb excitation experiments in $N = 88$ ^{144}Ba and $N = 90$ ^{146}Ba nuclei give direct evidence of enhanced octupole collectivity.

The experimental determination of octupole moments in ^{224}Ra, ^{226}Ra and ^{144}Ba, ^{146}Ba imply enhanced octupole collectivity in these nuclei but it does not differentiate between dynamic and static octupole collectivity. In order to be able to distinguish between these two modes, additional or supplementary experimental information is required. This is obtained from the behavior of energy levels in these nuclei. A discussion on this aspect is given in Sec. 4.6. In the same section, the available information on nuclear shape in these nuclei is also discussed.

5.3. Electric Dipole Moments and Pear-Shaped Nuclei

In this section, only a peripheral treatment of the topic will be provided. A full treatment is beyond the scope of this monograph. The reader is referred to a number of recent review articles for an in-depth coverage on electric dipole moments of elementary particles, atoms, molecules and nuclei [12–16].

The static electric dipole moment (EDM) of a particle or nucleus cannot exist unless simultaneously the invariances of both parity (P) and time-reversal (T) are violated and by the CPT-theorem, the charge-parity (CP) symmetry is violated. Therefore, an observation of a nonzero EDM with required sensitivity will confirm the CP violation.

For the particle or nucleus, the spin angular momentum **J** is the only vector to define a direction. The EDM **d** would be collinear with **J** because all components perpendicular to **J** would average to zero. The alignment of spin and EDM is what leads to violations of P and T.

Let us also follow the argument in [17] and the illustration given in Fig. 5.2. The EDM must vanish if there is invariance under parity transformation (P) for which $\mathbf{r} \rightarrow -\mathbf{r}$ or under the time-reversal transformation (T) for which $t \rightarrow -t$. Since, as mentioned, the orientation of the particle can be specified only by the orientation of its angular momentum **J**, the EDM **d** and **J** must transform their signs the same way under P and T invariance. Since, as shown in the figure (top right), **d** changes sign under P whereas **J** does

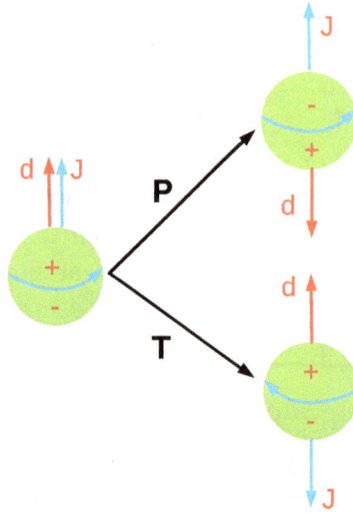

Fig. 5.2. Illustration of a spinning elementary particle or atom with spin orientation **J** and electric dipole moment (EDM) **d** [19]. Inversion through the origin or parity P, turns the particle as shown on top right. The spin orientation **J** remains the same but the sign of the EDM **d** is reversed. Time reversal T operation turns the particle as shown on bottom right. The EDM direction **d** remains the same as shown in the figure on the left but that of spin **J** is reversed. If the particle has an EDM **d**, the above mentioned two symmetries are violated (parity violation and time reversal invariance violation or CP violation). (Figure reproduced with permission from Professor Peter Butler.)

not, so **d** must vanish if there is P symmetry. Likewise, **d** does not change sign under T but **J** does (figure bottom right), so **d** must vanish if there is T symmetry. The observation of non-zero permanent EDM of a particle or nucleus would imply violation of both P and T symmetries. Assuming CPT invariance (CPT theorem), if T is violated then CP must also be violated.

An atom can possess an EDM due to the possible existence of (a) the electron EDM (d_e), (b) P- and T-violating electron-nucleus interactions and (c) hadronic CP violation [13]. Uptil now no EDM has been measured up to the level of experimental sensitivity and so the EDMs must be extremely small.

In the limit of a point-like nucleus and non-relativistic electrons, the EDM of the nucleus is completely screened by the atomic electrons so that the net atomic EDM vanishes. This is the Schiff theorem [18]. The nuclear EDM causes the orbital electrons to rearrange themselves to develop an electronic EDM equal and opposite to that of the nucleus [14]. In actual nuclei, none of the above limits hold fully. The nucleus has a finite size. The screening by atomic electrons is imperfect due to the finite size of the nucleus and relativistic electron motion. EDM of closed-shell (diamagnetic) atoms arises primarily from hadronic CP violation, i.e., CP violating interactions in the nuclear medium. The part of the nuclear EDM that survives screening by atomic electrons is characterized by nuclear Schiff moment which is defined as a mean square radius of the dipole charge distribution with the contribution of the centre of charge subtracted. It is the Schiff moment that directly induces an atomic EDM. It may be mentioned here that as a consequence of rotational invariance, the existence of nonzero Schiff moment requires nonzero nuclear spin **J**.

In open-shell (paramagnetic) atoms the EDM is mainly from electron EDM and P- and T-violating electron-nucleus interactions. In the discussion to follow, we will focus on the EDM of diamagnetic atoms.

We consider here the nuclear structure (nuclear shape) in the high Z nuclei as the effect produced by the Schiff moment increases faster than Z^2 [20]. At low spins, the nuclear shape could be axially symmetric and reflection-symmetric or axially symmetric and *reflection-asymmetric* octupole vibrational or stable pear-shaped octupole deformed. Considering the collective quadrupole-octupole deformed odd-A light actinide nuclei, in theoretical calculations of nuclear Schiff moments [21–23], it became clear that the P- and T-odd Schiff moments produced due to the parity and time invariance violating forces are typically enhanced by about two orders of magnitude compared to that in reflection symmetric deformed nuclei. Octupole deformation in these nuclei is manifested in the existence of low-lying parity doublet bands and parity doublet states. With the approximation that the octupole deformation is rigid, the ground state and its opposite parity low energy parity doublet partner state are projections onto good parity and angular momentum of the same "intrinsic state". The Schiff moment S is related to the operator corresponding to the electric dipole distribution weighted by radius squared, $\hat{S}_z$, the energy splitting of the parity doublet ΔE and the operator corresponding to the P- and T-violating nucleon-nucleon interaction $\hat{V}_{PT}$ in the expression below [24, 25]:

$$S \rightarrow -2\frac{J}{J+1}\frac{\langle\hat{S}_z\rangle\langle\hat{V}_{PT}\rangle}{\Delta E} \qquad (5.1)$$

where J is the ground-state spin and the brackets indicate expectation values in the intrinsic state. The small ΔE in the denominator is in part the reason for the enhancement of the Schiff moment.

The other factor in enhancement is the intrinsic-state expectation value $\langle\hat{S}_z\rangle$. It is generated by the collective quadrupole and octupole deformation of the entire nucleus. The octupole deformation enhances $\hat{S}_z$. The interaction expectation value $\langle\hat{V}_{PT}\rangle$ is harder to estimate [24, 26]. It can be inferred that the enhanced intrinsic Schiff moments and small energy difference in the denominator for atoms, the nuclei of which are octupole deformed, can provide sensitive tests of P, T violation in the nucleon-nucleon interactions. In [27], it was found that the contribution to the Schiff moment is nearly the same from soft collective octupole vibrations and static octupole deformation.

Let us consider odd-A nuclei in the light actinide mass region which are octupole deformed and are favorable for atomic electric dipole moment (EDM) measurements as these have strong octupole correlations. Table 5.1 gives the details on low energy parity doublets in some of these nuclei — spin-parity J^π for the ground state and the excited state members and their energy difference ΔE:

Out of the above list, ^{225}Ra is most interesting for experimental measurements of atomic EDM. In a reanalysis of the then existing experimental data on ^{225}Ra in [29], evidence of reflection asymmetric shape was found with the existence of low-lying parity doublet bands. In the $K^\pi = 1/2^\pm$ parity doublet bands, the energy difference between the $J = 1/2^+$ ground state and the $J = 1/2^-$ parity doublet state is low ($\Delta E = 55.16\,\text{keV}$). Since the ground state spin of ^{225}Ra is $J = 1/2$, it minimizes the effect of stray quadrupole

Table 5.1. Spin-parity J^π and energy difference ΔE for low energy parity doublets in some of the odd-A light actinide nuclei. Unless otherwise noted, the J^π and ΔE values are from [10].

Nucleus	$J^\pi_{\text{g.s.}}$ - J^π_{ex}	ΔE(keV)
$^{223}\text{Ra}_{135}$	$3/2^+ - 3/2^-$	50.128
$^{225}\text{Ra}_{137}$	$1/2^+ - (1/2^-)$	55.16
$^{223}\text{Ac}_{134}$	$(5/2^-) - (5/2^+)$	64.62
$^{225}\text{Ac}_{136}$	$(3/2^-) - (3/2^+)$	40.09
$^{229}\text{Pa}_{138}$	$5/2^+ - 5/2^{-*}$	$(60 \pm 50)\,\text{eV}^*$

*From [28].

electric fields. Although ^{225}Ra in ground state is radioactive, it has a long halflife of 14.9 days suitable for experimentation and has a favourable associated atomic structure. Recent experimental measurements [30] found the limit of atomic EDM in ^{225}Ra to be less than 1.4×10^{-23} e cm with 95% confidence upper limit. The details of the measurement method can be found in the reference [30].

References

1. B.A. Robson, *J. H.E. Phys.: Gravitation Cosmology* **4**, 166 (2018).
2. N.E. Mavromatos, *J. Phys: Conf. Series* **952**, 012006 (2018).
3. T.D. Lee and C.N. Yang, *Phys. Rev.* **104**, 254 (1956).
4. C.S. Wu *et al.*, *Phys. Rev.* **105**, 1413 (1957).
5. J.H. Christenson *et al.*, *Phys. Rev. Lett.* **13**, 138 (1964).
6. M. Kobayashi and T. Maskawa, *Prog. Theor. Phys.* **49**, 652 (1973).
7. B. Aubert *et al.*, (Babar collaboration), *Phys. Rev. Lett.* **87**, 091801 (2001).
8. K. Abe *et al.*, (Belle collaboration), *Phys. Rev. Lett.* **87**, 091802 (2001).
9. R. Aaij *et al.*, (LHCb collaboration), *Phys. Rev. Lett.* **122**, 211803 (2019).
10. Brookhaven National Data Center, ENSDF files; http://www.nndc.bnl.gov
11. W. Nazarewicz *et al.*, *Nucl. Phys. A* **467**, 437 (1987).
12. T.E. Chupp *et al.*, *Rev. Mod. Phys.* **91**, 015001 (2019).
13. N. Yamanaka *et al.*, *Eur. Phys. J. A* **53**, 54 (2017).
14. J. Engel *et al.*, *Prog. Part. Nucl. Phys.* **71**, 21 (2013).
15. E. D. Commins, *J. Phys. Soc. Japan* **76**, 111010 (2007).
16. N. Fortson *et al.*, *Physics Today* **56**, 33 (2003).
17. N.F. Ramsey, *Annu. Rev. Nucl. Part. Sci.* **40**, 1 (1990).
18. L.I. Schiff, *Phys. Rev.* **132**, 2194 (1963).

19. Talk on *Pear-shaped Nuclei and CP-Violation* by P.A. Butler, SPES — Nusprasen Workshop, Pisa 1–2 Feb. 2018.
20. V.V. Flambaum, *Phys. Rev. A* **77**, 024501 (2008).
21. V. Spevak *et al.*, *Phys. Lett. B* **359**, 254 (1995).
22. V. Spevak *et al.*, *Phys. Rev. C* **56**, 1357 (1997).
23. J. Engel *et al.*, *Phys. Rev. C* **61**, 035502 (2000).
24. J. Engel *et al.*, *Phys. Rev. C* **68**, 025501 (2003).
25. P.A. Butler, *J. Phys. G: Nucl. Part. Phys.* **43**, 073002 (2016).
26. J. Dobaczewski and J. Engel, *Phys. Rev. Lett.* **94**, 232502 (2005).
27. V.V. Flambaum and V.G. Zelevinsky, *Phys. Rev. C* **68**, 035502 (2003).
28. I. Ahmad *et al.*, *Phys. Rev. C* **92**, 024313 (2015).
29. R.K. Sheline *et al.*, *Phys. Lett. B* **219**, 47 (1989).
30. M. Bishof *et al.*, *Phys. Rev. C* **94**, 025501 (2016).

Partial List of Review Articles and Review-Like Papers on Reflection Asymmetric Nuclei

1. **Octupole collectivity in nuclei**, P.A. Butler, *J. Phys. G: Nucl. Part. Phys.* **43**, 073002 (2016).
2. Studies of the shapes of heavy pear-shaped nuclei at ISOLDE, P.A. Butler, *AIP Conference Proceedings* **1753**, 030002 (2016).
3. Octupole deformation in the ground states of even-even nuclei: A global analysis within the covarient density functional theory, S.E. Agbemava *et al.*, *Phys. Rev. C* **93**, 044304 (2016).
4. Developments in the studies of pear-shaped nuclei and their impact on searches for C-P violation in atoms, P. A. Butler and L. Willmann, *Nuclear Physics News* **25**, 12 (2015).
5. Pear shaped nuclei, A.A. Raduta in *Nuclear Structure with Coherent States* (Springer, 2015) p. 235.
6. Pear-shaped nuclei: Nuclear models and the standard model, P.A. Butler, *Acta Phys. Pol. B* **45**, 127 (2014).
7. Cross-over between different symmetries, S. Frauendorf in *Proceedings, 5th International Conference on Fission and Properties of Neutron Rich Nuclei (ICFN5)*: Sanibel Island, Florida, USA, November 4–10, 2012, eds., J.H. Hamilton and A.V. Ramayya, (World Scientific, 2013) p. 39.
8. New results on octupole collectivity, M.P. Carpenter *et al.*, *Journal of Physics, Conf. Series* **312**, 092006 (2011).

9. Global systematics of octupole excitations in even-even nuclei, L.M. Robledo and G.F. Bertsch, *Phys. Rev. C* **84**, 054302 (2011).

10. Reduced electric-octupole transition probabilities, $B(E3; 0_1^+ \rightarrow 3_1^-)$ — An update T. Kibédi and R.H. Spear, *At. Data Nucl. Data Tables* **80**, 35 (2002).

11. **Fast nuclear rotation and octupole deformation**, W. Urban *et al.*, *Acta Phys. Pol.* **32**, 2527 (2001).

12. Spontaneous symmetry breaking in rotating nuclei, S. Frauendorf, *Rev. Mod. Phys.* **73**, 463 (2001).

13. Octupole shapes, P.A. Butler, *Phys. Scr. T* **88**, 7 (2000).

14. Intrinsic reflection asymmetry in nuclei, P.A. Butler, *Acta Phys. Pol. B* **29**, 289 (1998).

15. Nuclear Pears: Recent developments and future prospects, P.A. Butler *et al.*, in *Heavy Ion Physics* **7**, 1 (1998).

16. **Intrinsic reflection asymmetry in atomic nuclei**, P.A. Butler and W. Nazarewicz, *Rev. Mod. Phys.* **68**, 349 (1996).

17. **Octupole shapes in nuclei**, I. Ahmad and P. A. Butler, *Annu. Rev. Nucl. Part. Sci.* **43**, 71 (1993).

18. Octupole collectivity in nuclei deduced from Coulomb excitation measurements, P.A. Butler, *Acta Phys. Pol. B* **24**, 117 (1993).

19. Quadrupole and octupole shapes in nuclei, D. Cline, *Nucl. Phys. A* **557**, 615c (1993).

20. High — spin spectroscopy of reflection asymmetric nuclei, N. Schulz, *Acta Phys. Pol. B* **24**, 43 (1993).

21. Rotational bands in deformed odd-A nuclei in the actinide region, K. Jain and A. K. Jain, *At. Data Nucl. Data Tables* **50**, 269 (1992).

22. Octupole shapes and shape changes at high spins in the $Z \approx 58$, $N \approx 88$ nuclei, W. Nazarewicz and S.L. Tabor, *Phys. Rev. C* **45**, 2226 (1992).

23. A systematic study of the octupole correlations in the lanthanides with realistic forces, J.L. Egido and L.M. Robledo, *Nucl. Phys. A* **545**, 589 (1992).

24. Intrinsic dipole moments in reflection-asymmetric nuclei, P.A. Butler and W. Nazarewicz, *Nucl. Phys. A* **533**, 249 (1991).

25. Reflection-asymmetric shapes in odd-A actinide nuclei, S. Ćwiok and W. Nazarewicz, *Nucl. Phys. A* **529**, 95 (1991).

26. Low energy octupole and dipole modes in nuclei, W. Nazarewicz, *Nucl. Phys. A* **520**, 333c (1990).

27. Nuclear shapes in mean field theory, S. Åberg, H. Flocard and W. Nazarewicz, *Annu. Rev. Nucl. Part. Sci.* **40**, 439 (1990).

28. Intrinsic states of deformed odd-A nuclei in the mass regions (151 $\leq A \leq$ 193) and ($A \geq$ 221), A.K. Jain *et al.*, *Rev. Mod. Phys.* **62**, 393 (1990).

29. Information on octupole correlations in nuclei from γ-ray spectroscopy, P.A. Butler in *Heavy Ions in Nuclear and Atomic Physics*, ed., Z. Wilhelmi and G. Szeflińska (1989) (Bristol & Philadelphia: Adam Hilger), p. 295.

30. Reflection-asymmetric rotor model of odd A ~219 – 229 nuclei, G.A. Leander and Y.S. Chen, *Phys. Rev. C* **37**, 2744 (1988).

31. Octupole vibrations in nuclei, Stanislaw G. Rohoziński, *Rep. Prog. Phys.* **51**, 541 (1988).

32. Octupole shapes and shape changes at high spins in Ra and Th nuclei, W. Nazarewicz *et al.*, *Nucl. Phys. A* **467**, 437 (1987).

33. Low-energy collective $E1$ mode in nuclei, G.A. Leander *et al.*, *Nucl. Phys. A* **453**, 58 (1986).

34. Rotational consequences of stable octupole deformation in nuclei, W. Nazarewicz and P. Olanders, *Nucl. Phys. A* **441**, 420 (1985).

35. "Static" octupole deformation, G.A. Leander in *Nuclear Structure '85*, Proc. Niels Bohr Cent. Conf., Cpoenhagen, Eds. R.A. Broglia *et al.*, (North-Holland, 1985), p. 249.

36. "Static" octupole deformation at high spin, W. Nazarewicz in *Nuclear Structure '85, Proc. Niels Bohr Cent. Conf.*, Cpoenhagen, Eds. R.A. Broglia *et al.*, (North-Holland, 1985), p. 263.

37. Analysis of octupole instability in medium-mass and heavy nuclei, W. Nazarewicz *et al.*, *Nucl. Phys. A* **429**, 269 (1984).

38. Properties of the yrast states in the actinides, J.L. Egido and P. Ring, *Nucl. Phys. A* **423**, 93 (1984).

39. Intrinsic reflection asymmetry in odd-A nuclei, G. A. Leander and R.K. Sheline, *Nucl. Phys. A* **413**, 375 (1984).

40. Magnetic moments as a probe for rotational alignment, Y.S. Chen and S. Frauendorf, *Nucl. Phys. A* **393**, 135 (1983).

41. The breaking of intrinsic reflection symmetry in nuclear ground states, G.A. Leander *et al.*, *Nucl. Phys. A* **388**, 452 (1982).

Index